Haruto

Akari

Sara

Yu

Let's learn mathematics together!

Grade 1 Vol. 2

Dear Teachers and Parents

This textbook has been compiled in the hope that children will enjoy learning through acquiring mathematical knowledge and skills. The unit pages are carefully written to ensure that students can understand the content they are expected to master at that grade level. In addition, the "More Maths!" section at the end of the book is designed to ensure that each student has mastered the content of the main text, and is intended to be handled selectively according to the actual conditions and interests of each child. We hope that this textbook will help children develop an interest in mathematics and become more motivated to learn.

Infectious Disease Control
In this textbook, pictures of activities and illustrations of characters do not show children wearing masks, etc., in order to cultivate children's rich spirit of communicating and learning from each other. Please be careful to avoid infectious diseases when conducting learning activities.

QR codes are used to connect to Internet content by launching a QR code-reading application on a smartphone or tablet and reading the code with a camera.
The QR Code can be used to access content on the Internet.
The code can also be used at the address below.
https://r6.gakuto-plus.jp/s1a01

Note: This book is an English translation of a Japanese mathematics textbook. The only language used in the contents on the Internet is Japanese.

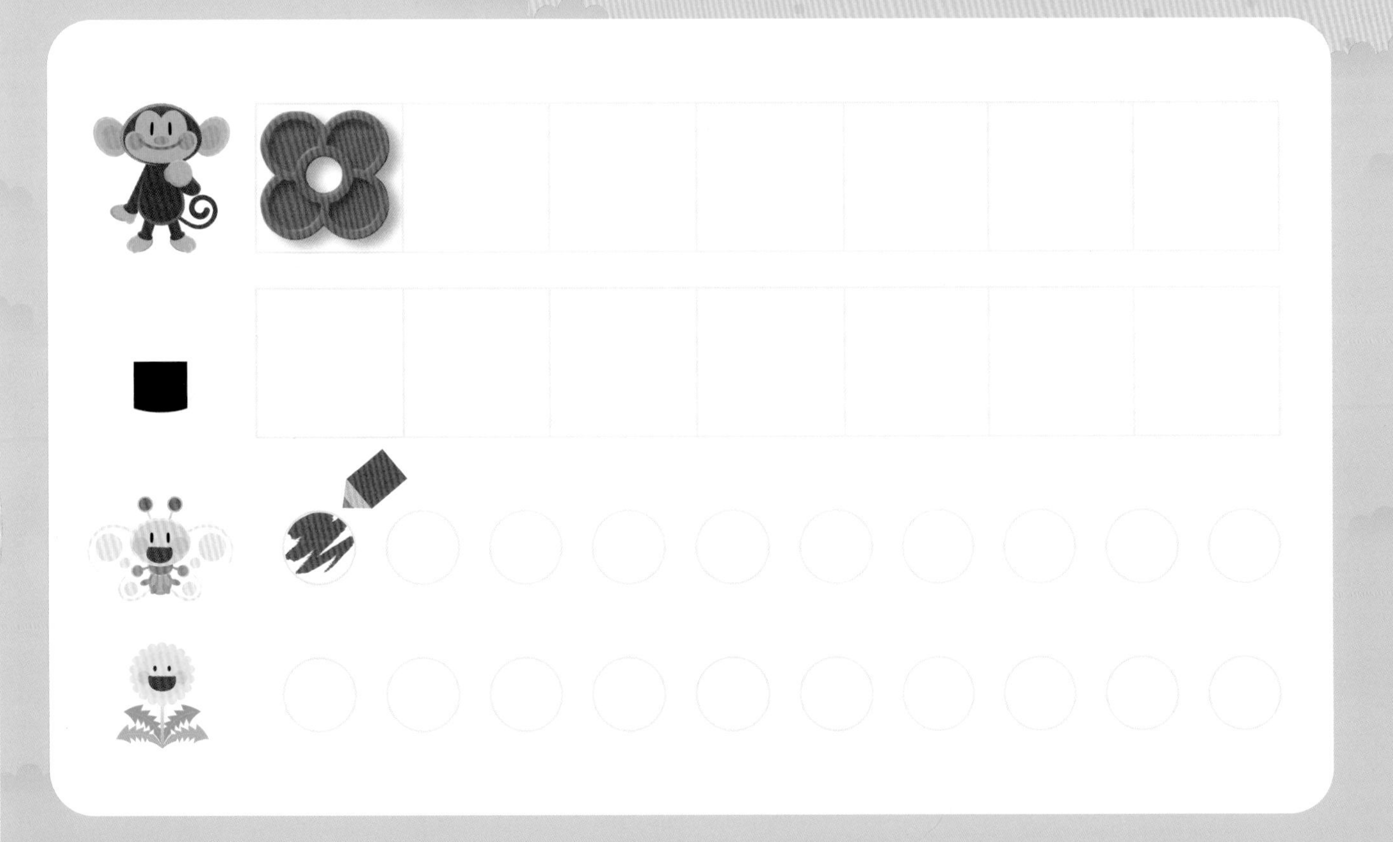

1
Numbers up to 10

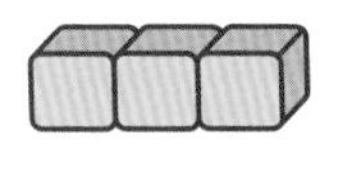

three

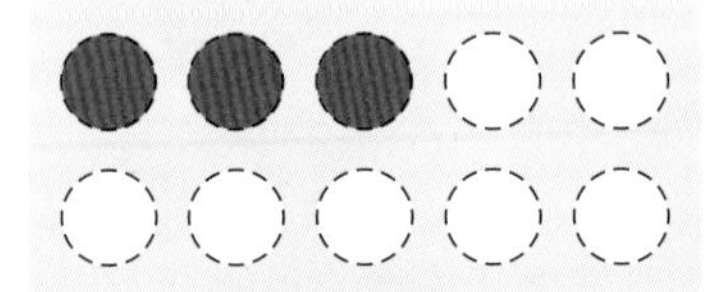

3

3

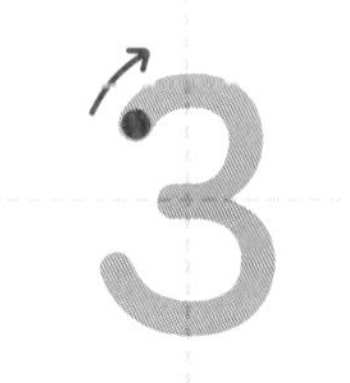

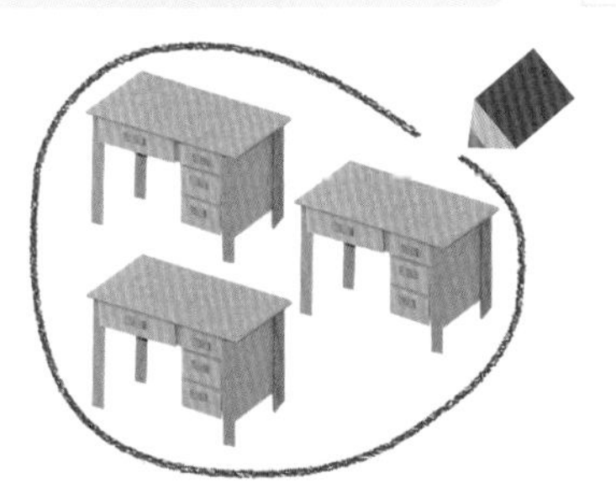

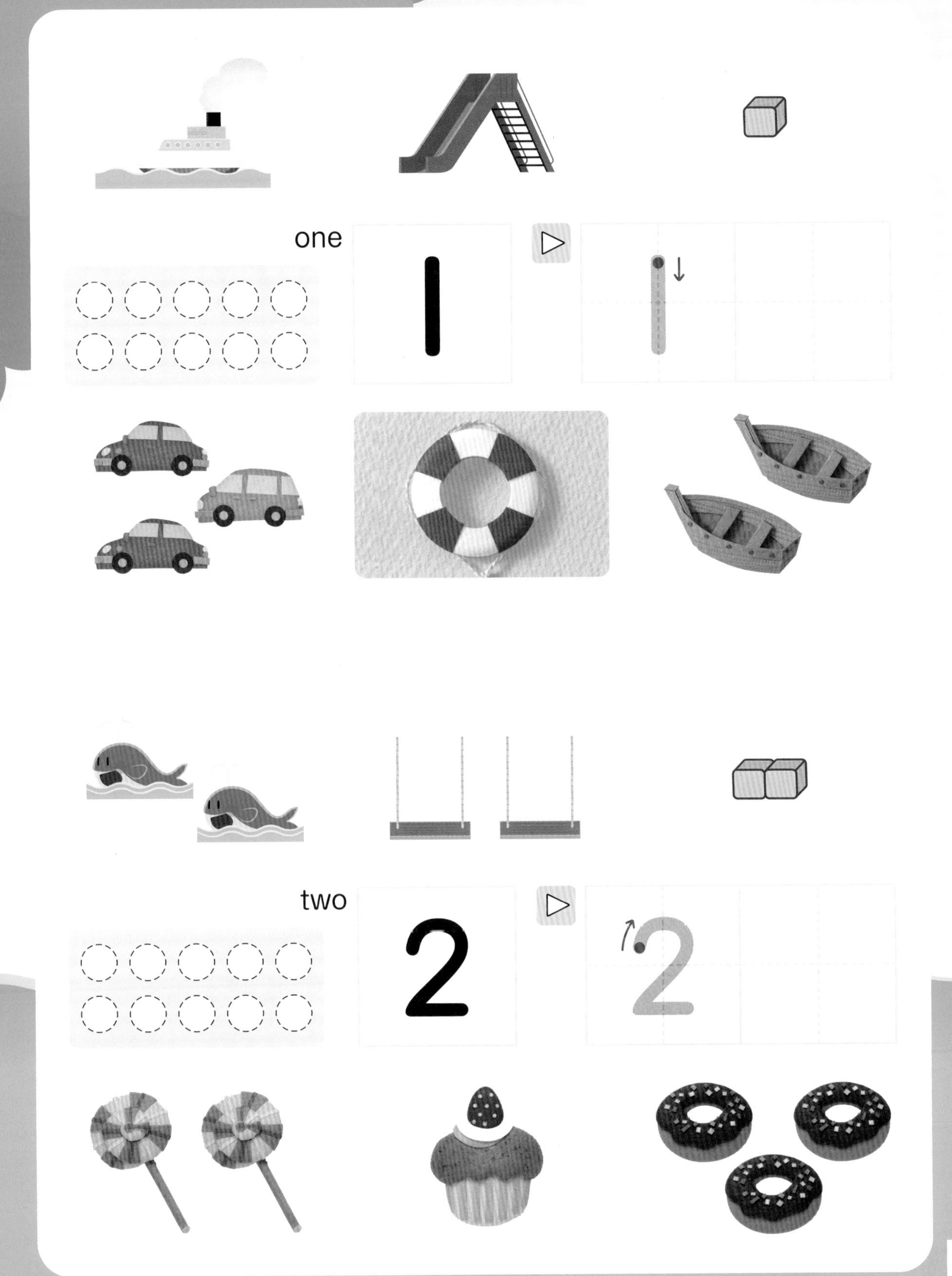
one
1
two
2

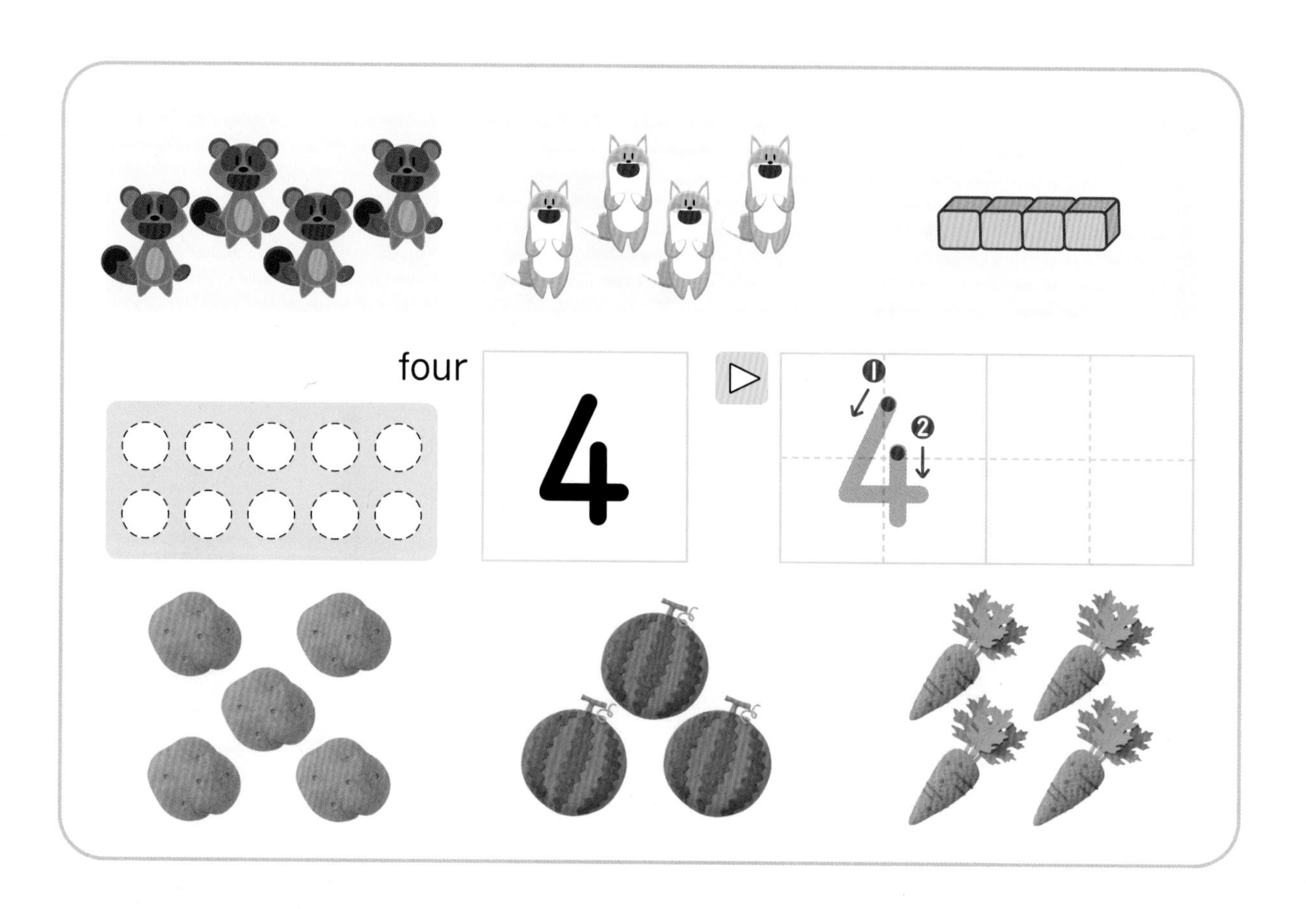
four
4
4

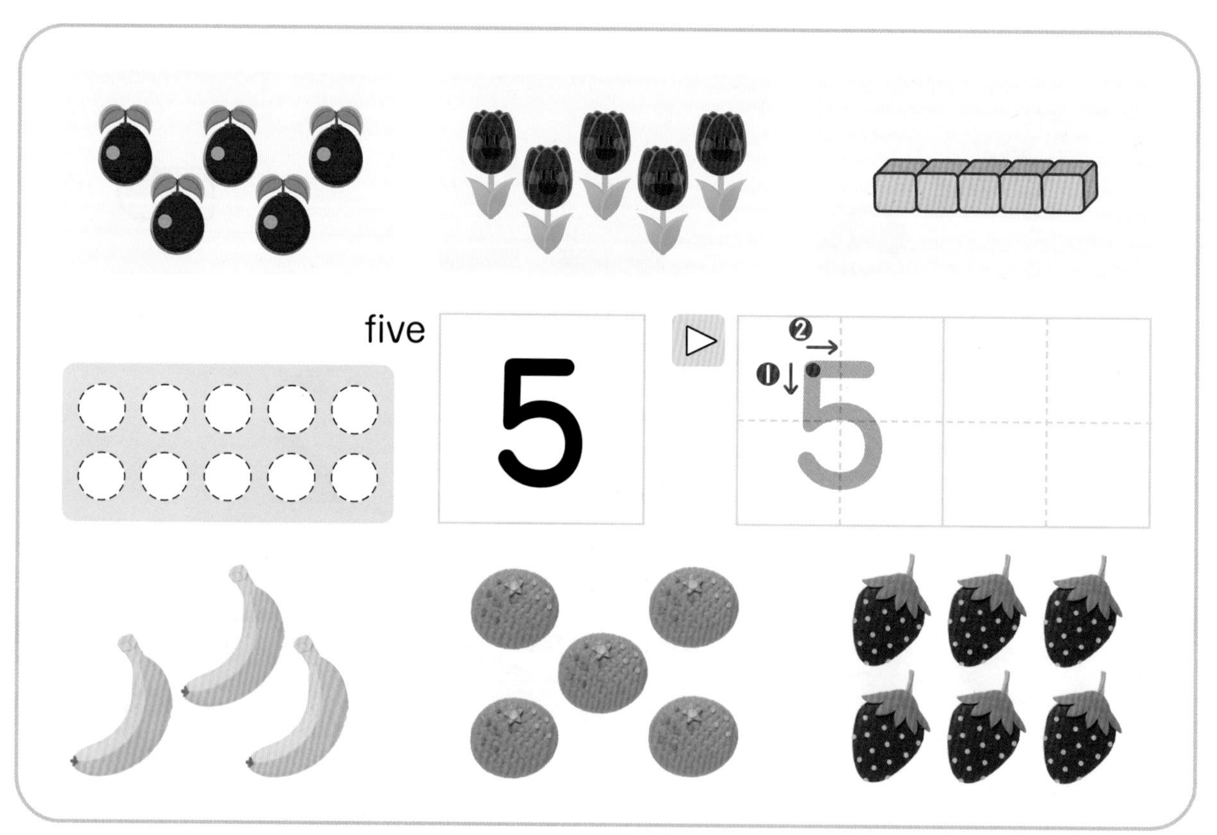
five
5
5

2
2
5
3
4
5
2

2 two
3
three

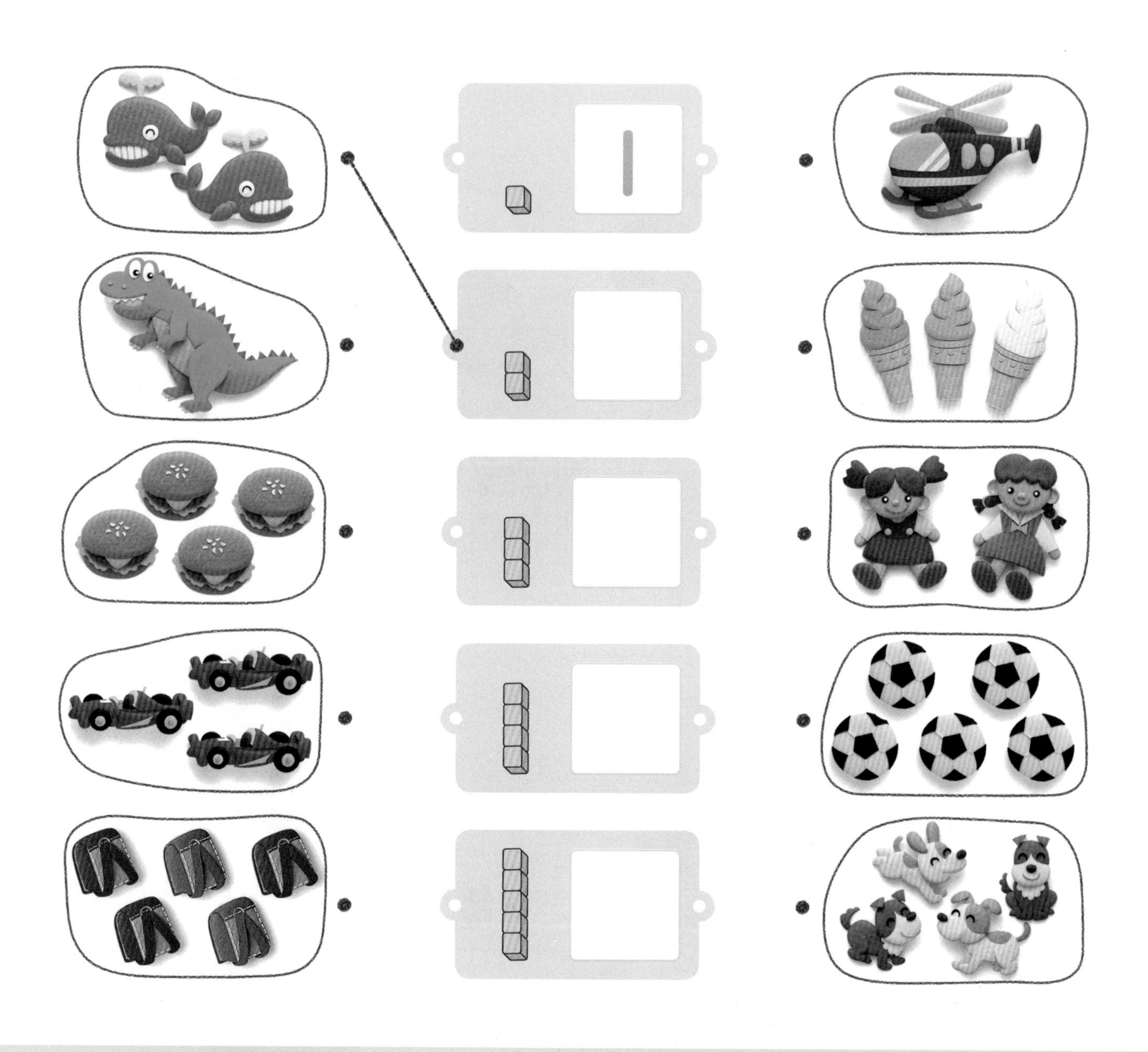
1

six
6
seven
7

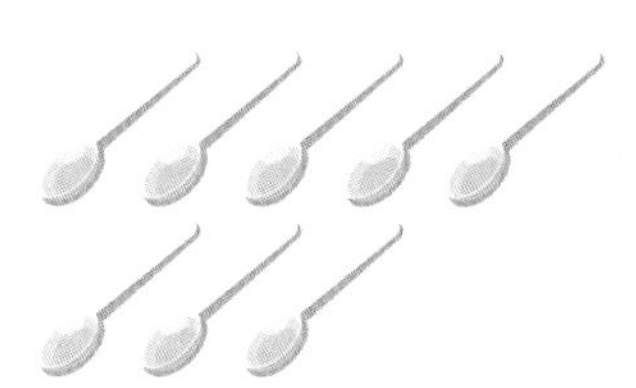
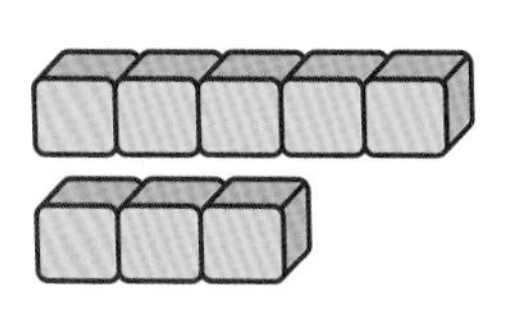

eight

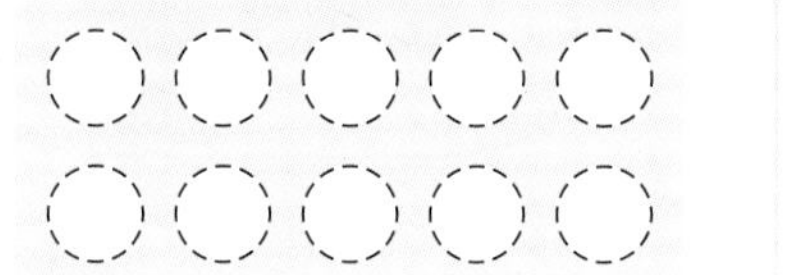

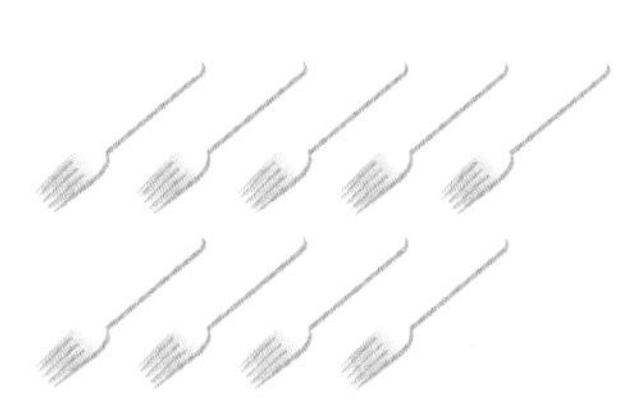

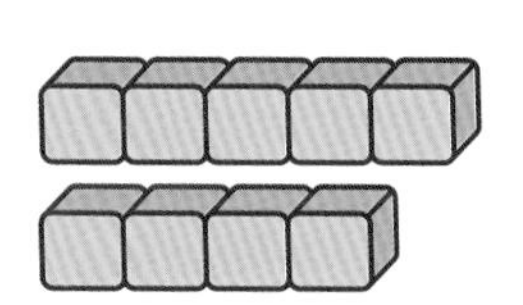

nine

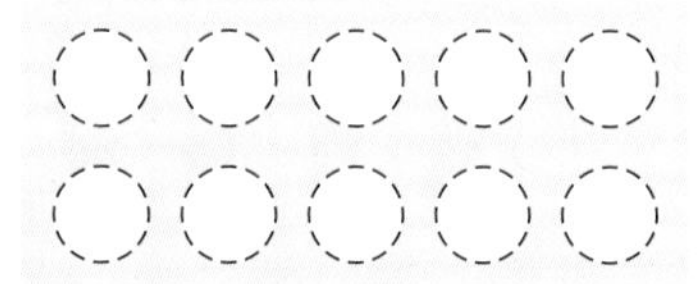

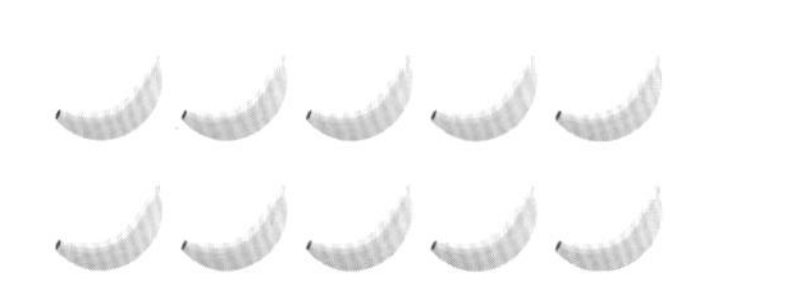

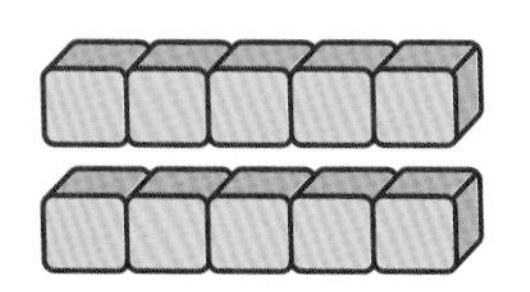

ten

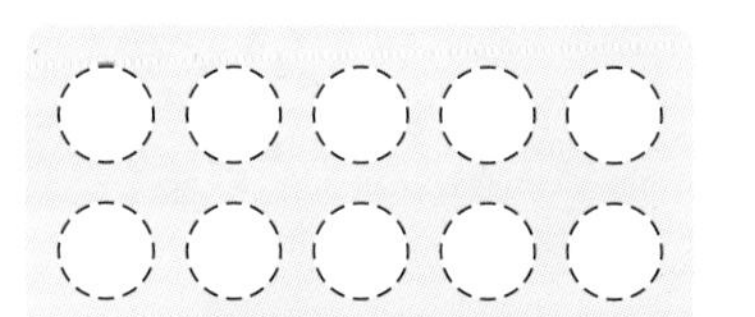
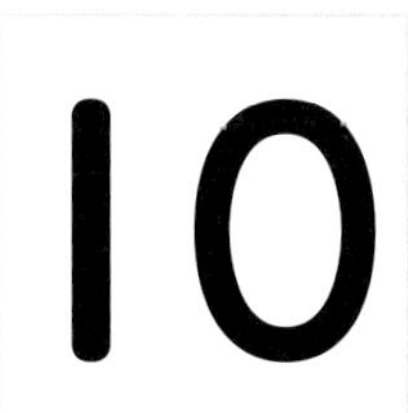

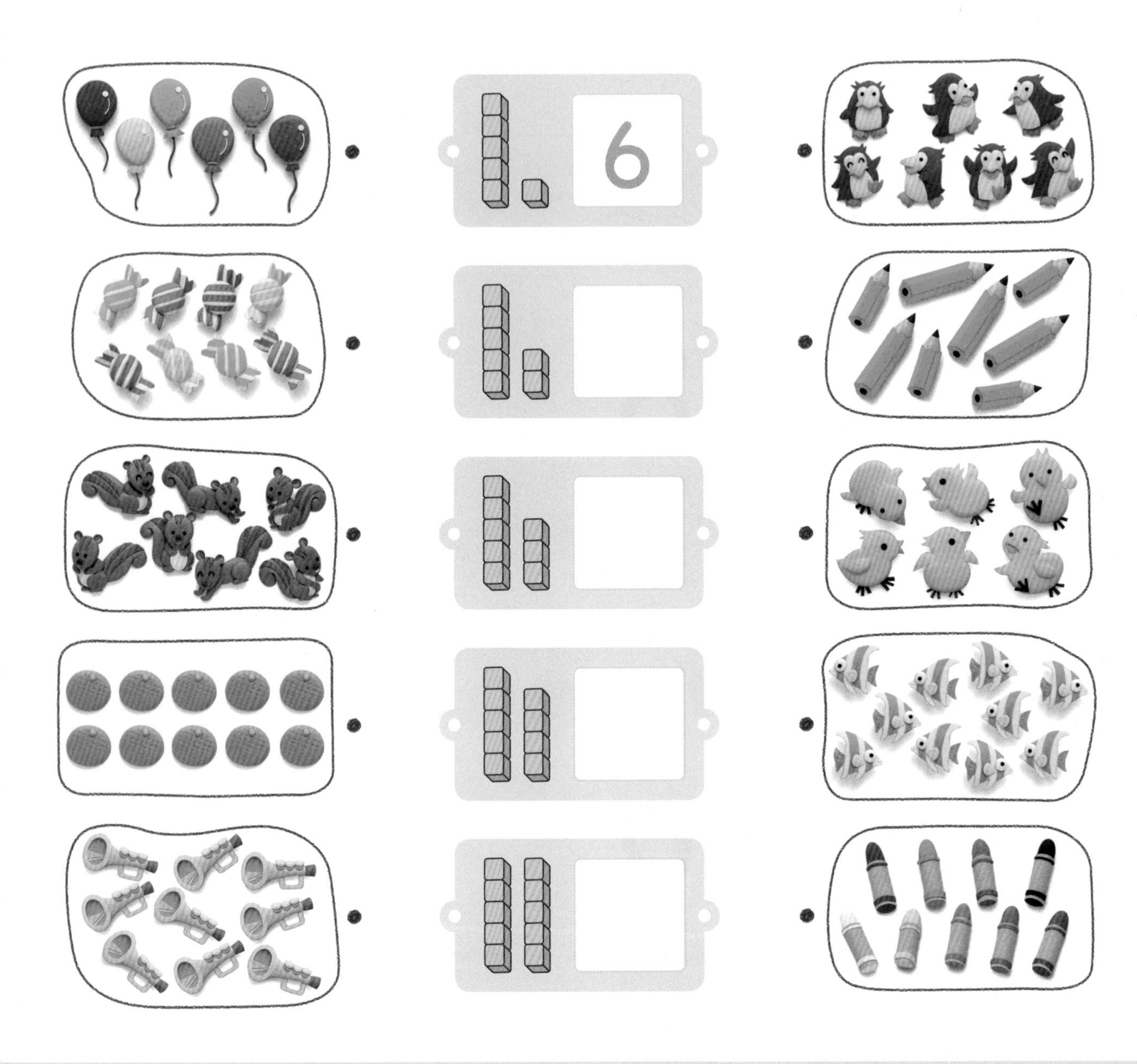
6

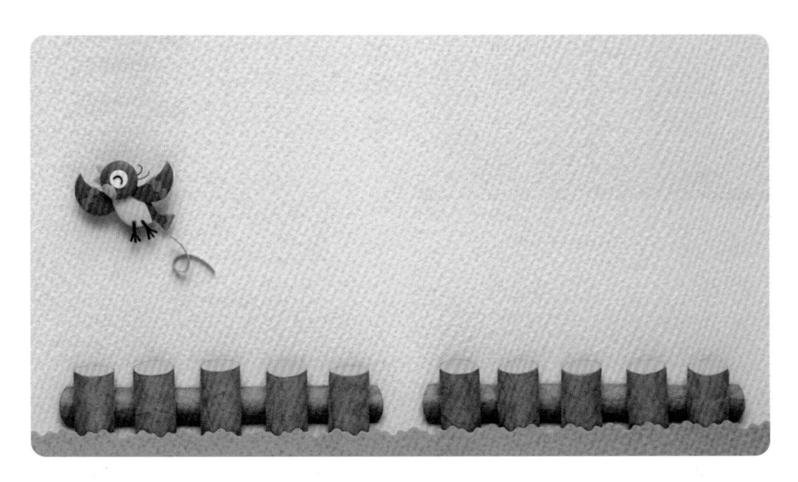

zero

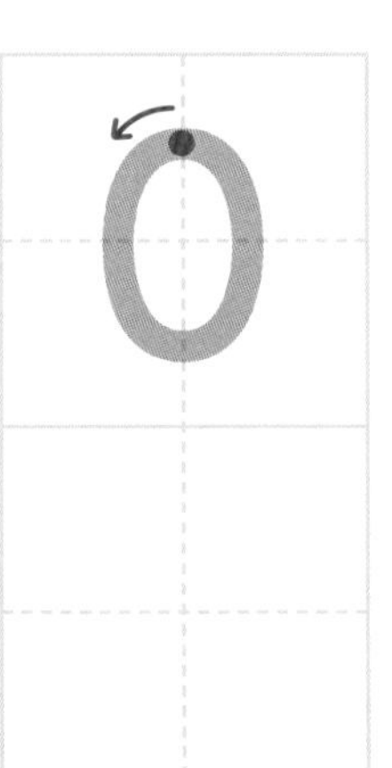

Which is more?

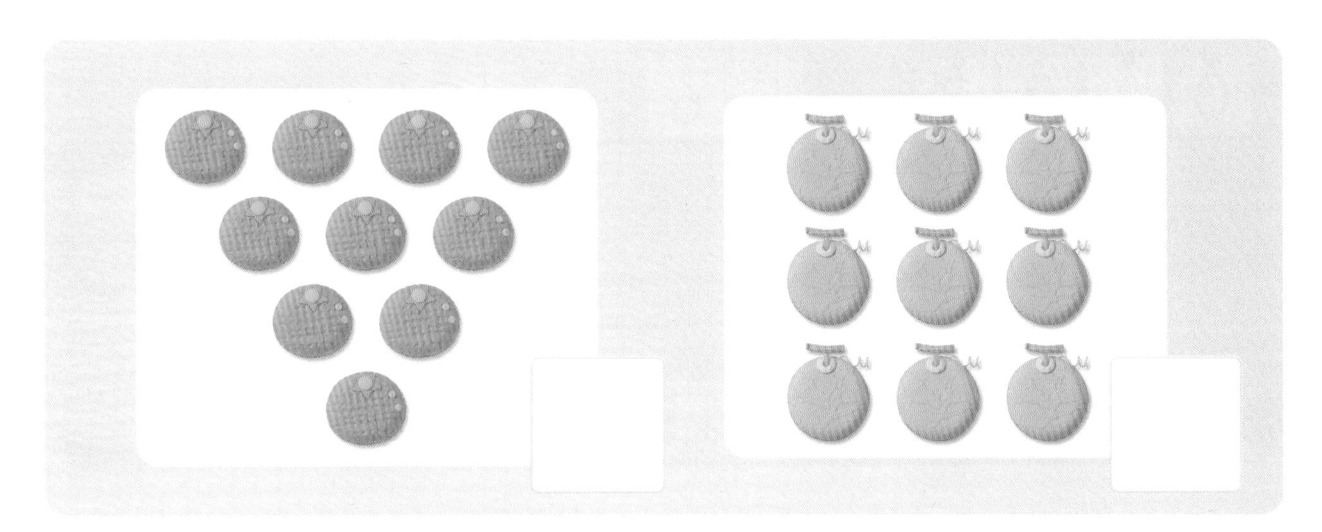

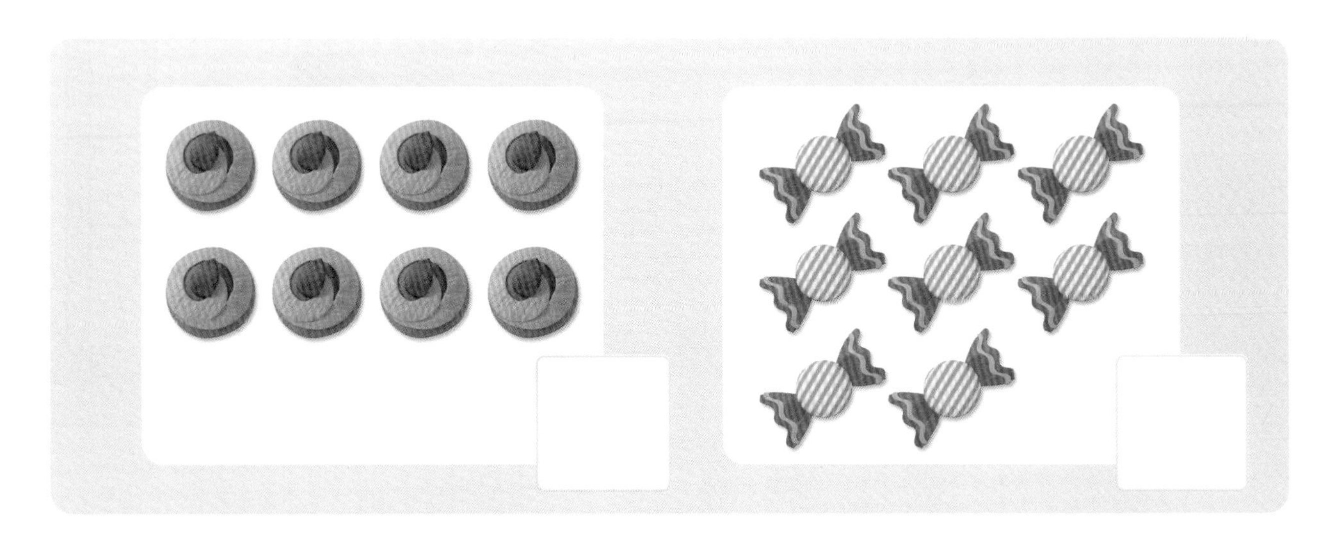

Which is larger?

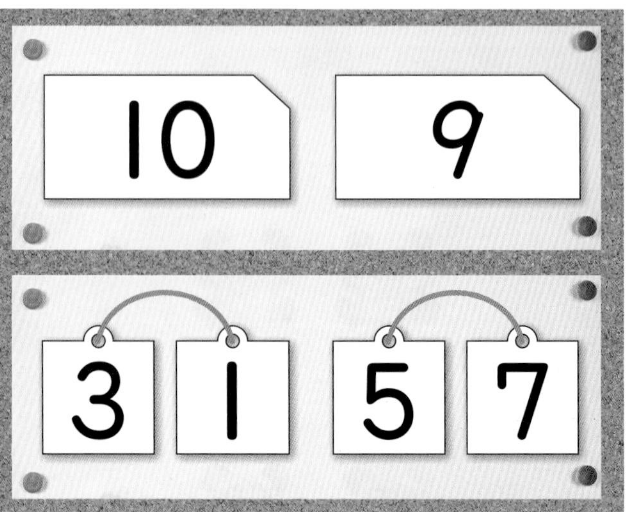

Ordering the cards

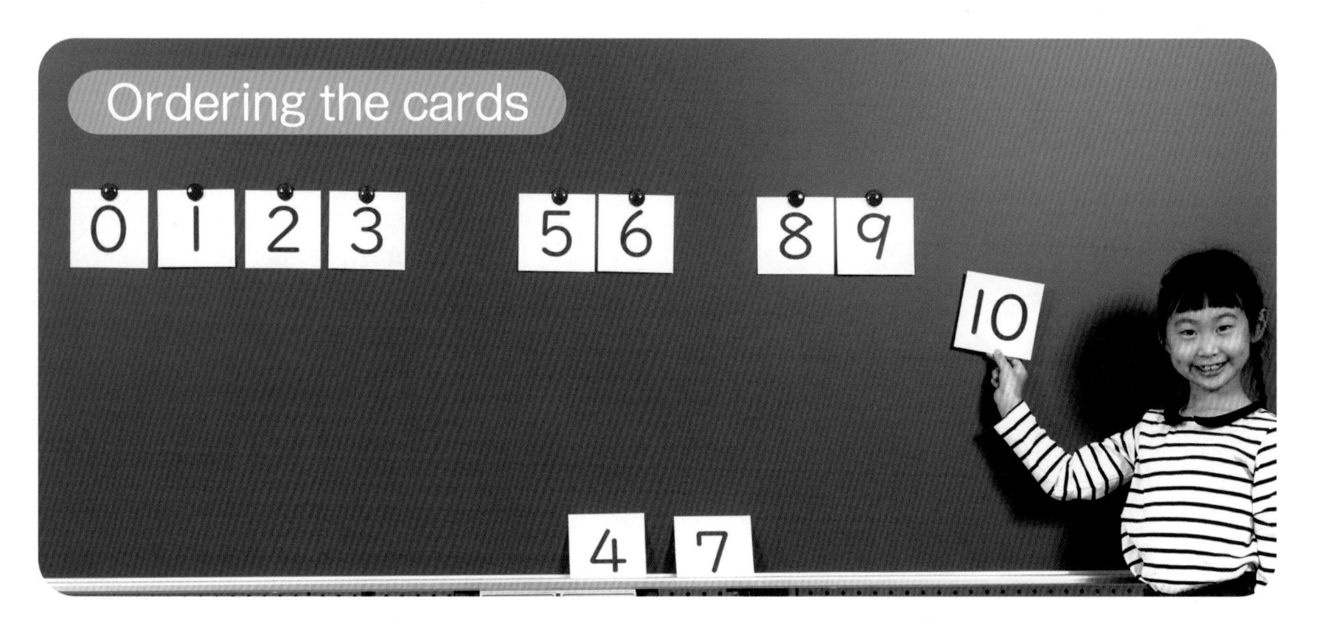

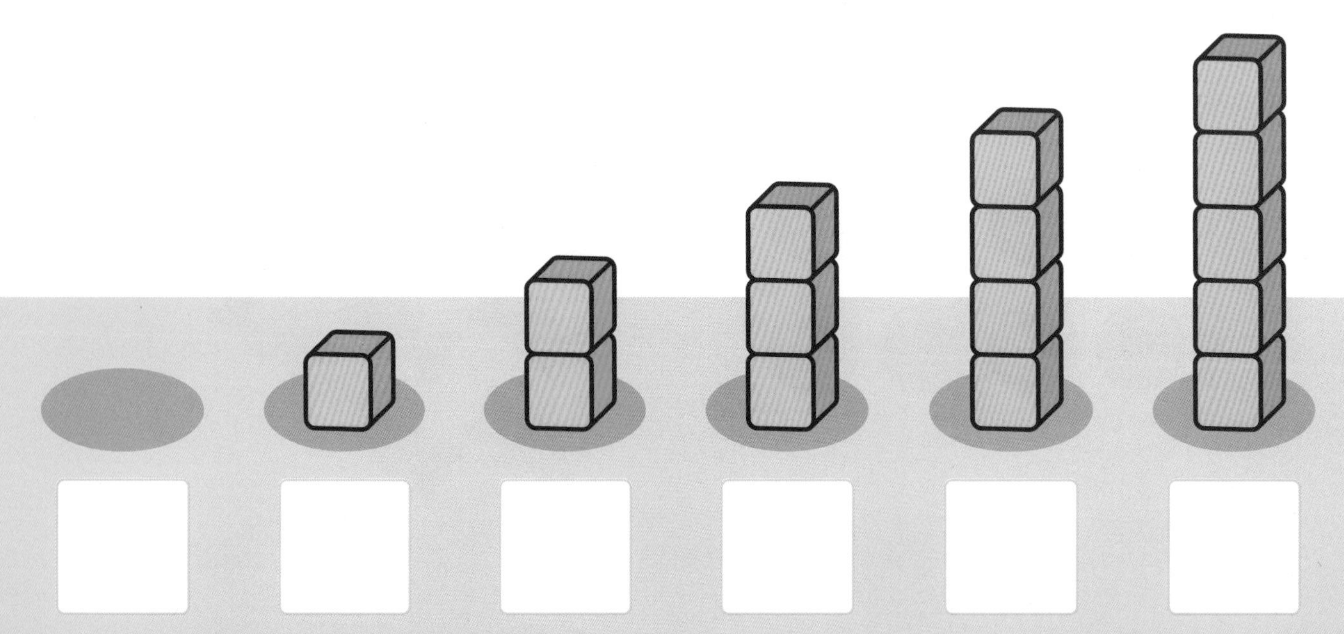

Supplementary Problems → p.89

2 Decomposing and Composing Numbers

Let's decompose 5→

5 balls are divided. How many balls are there in each side?

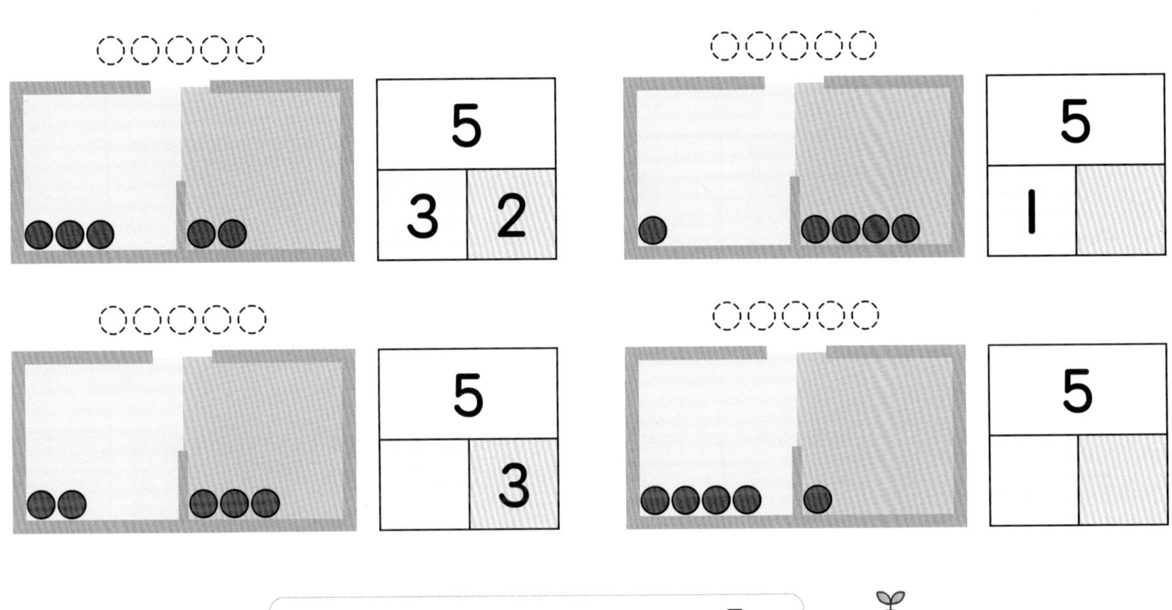

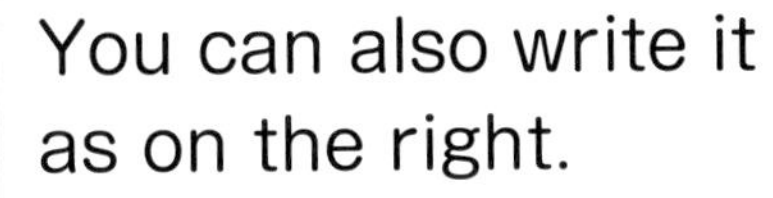

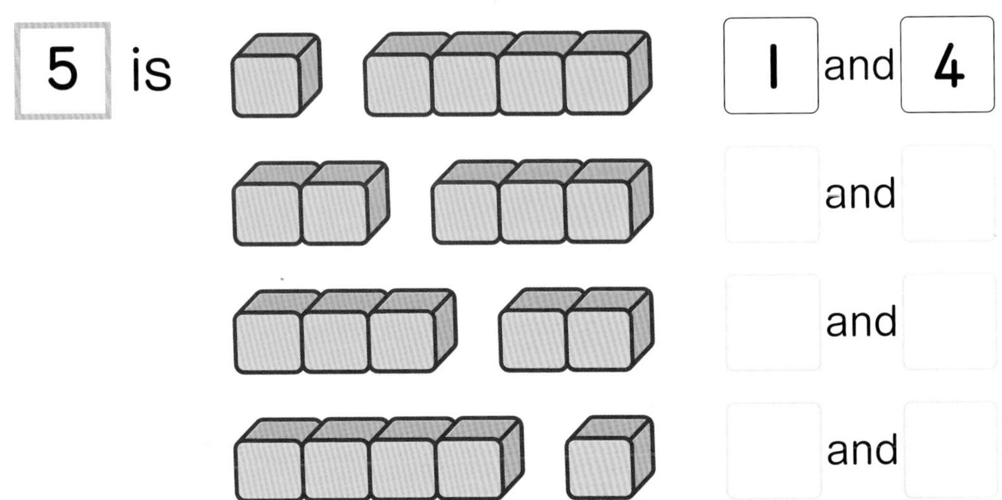

There are four ways to decompose 5.

6 balls are divided. How many balls are there in each side?

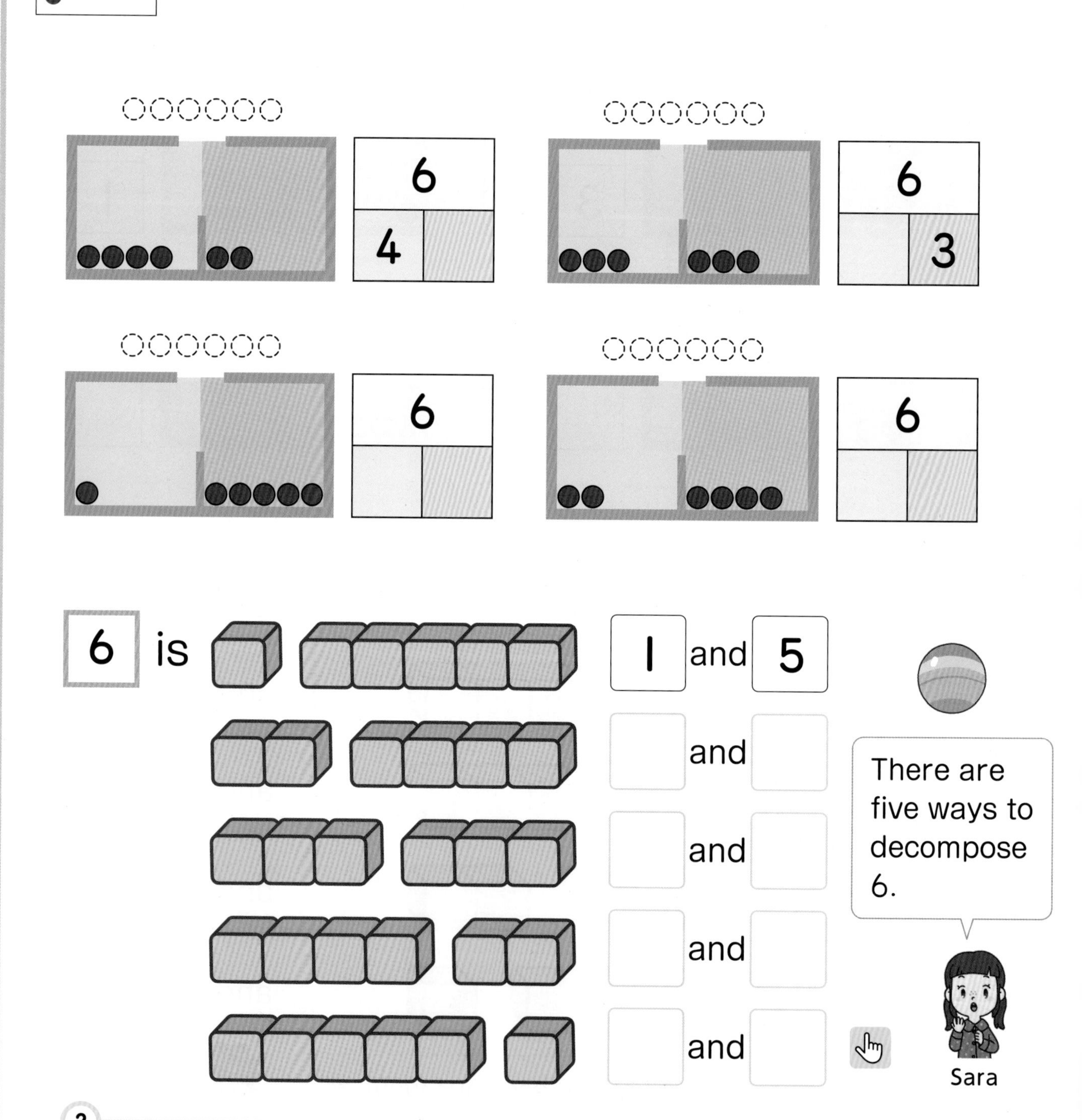

? Based on how you decomposed 6, how many ways are there to decompose 7 into two numbers?

7

I got 7 marbles.

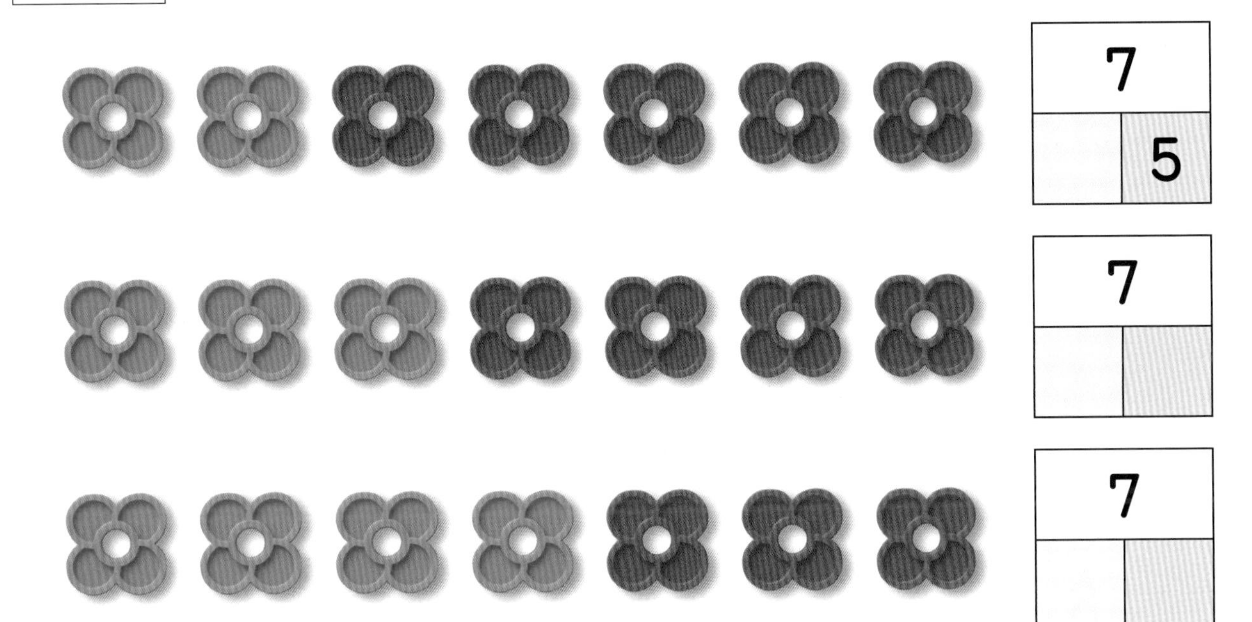

7	
	5

7	

7	

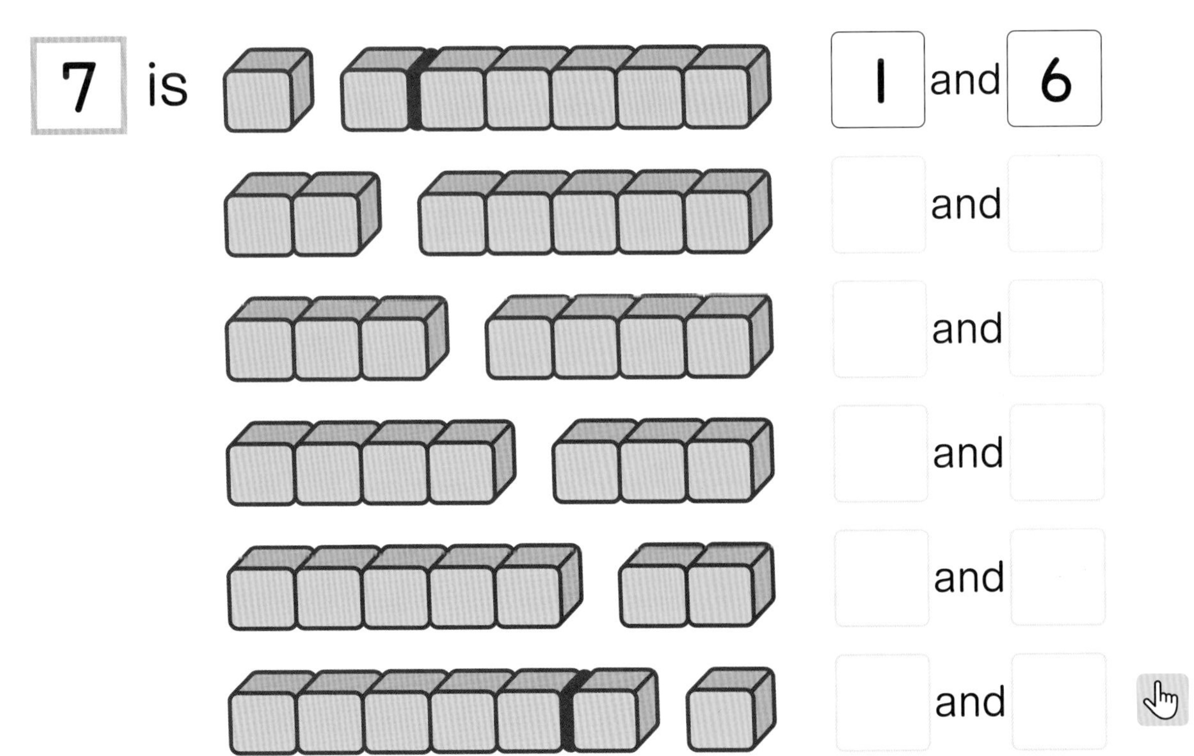

7 is 1 and 6

and

and

and

and

and

I have 8 marbles.

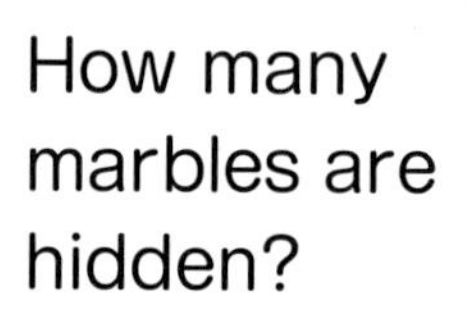

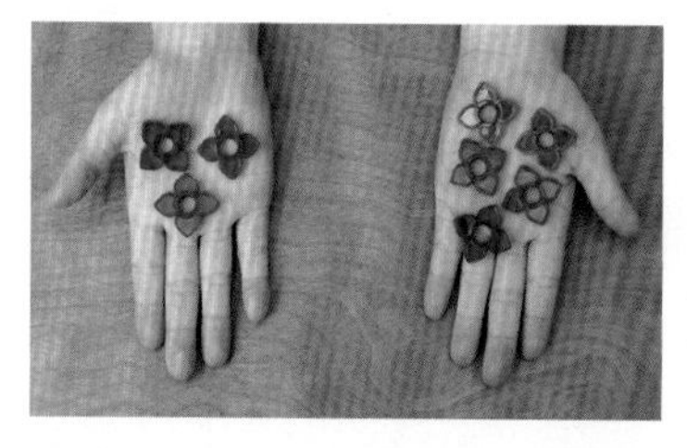

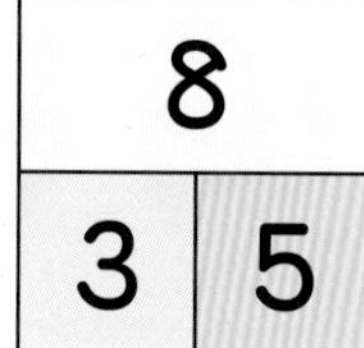

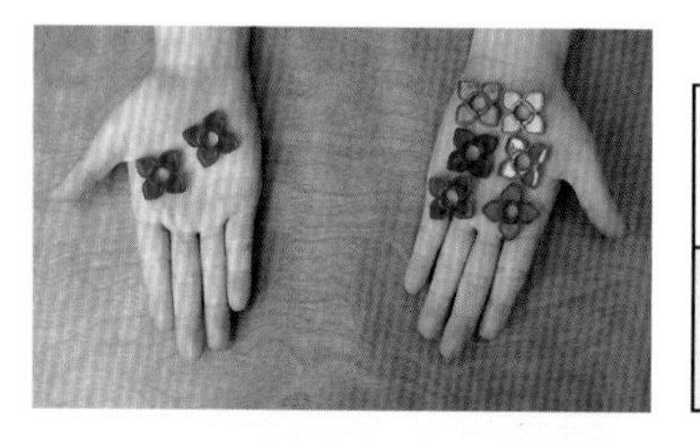

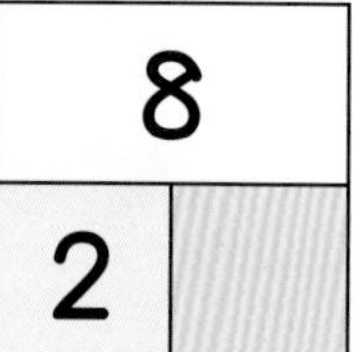

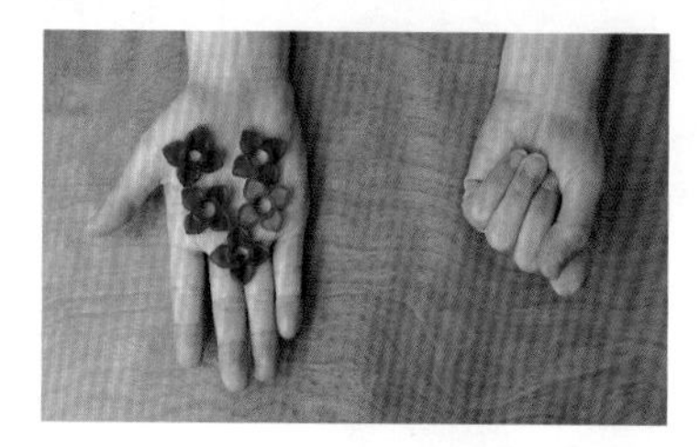

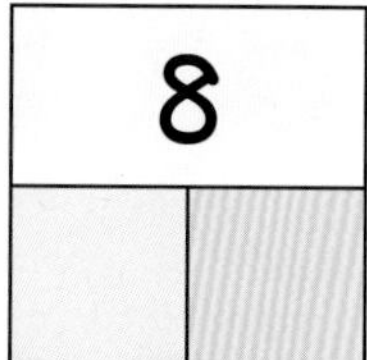

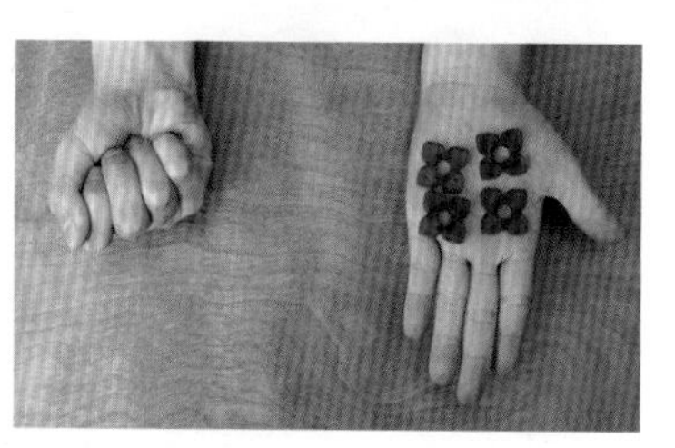

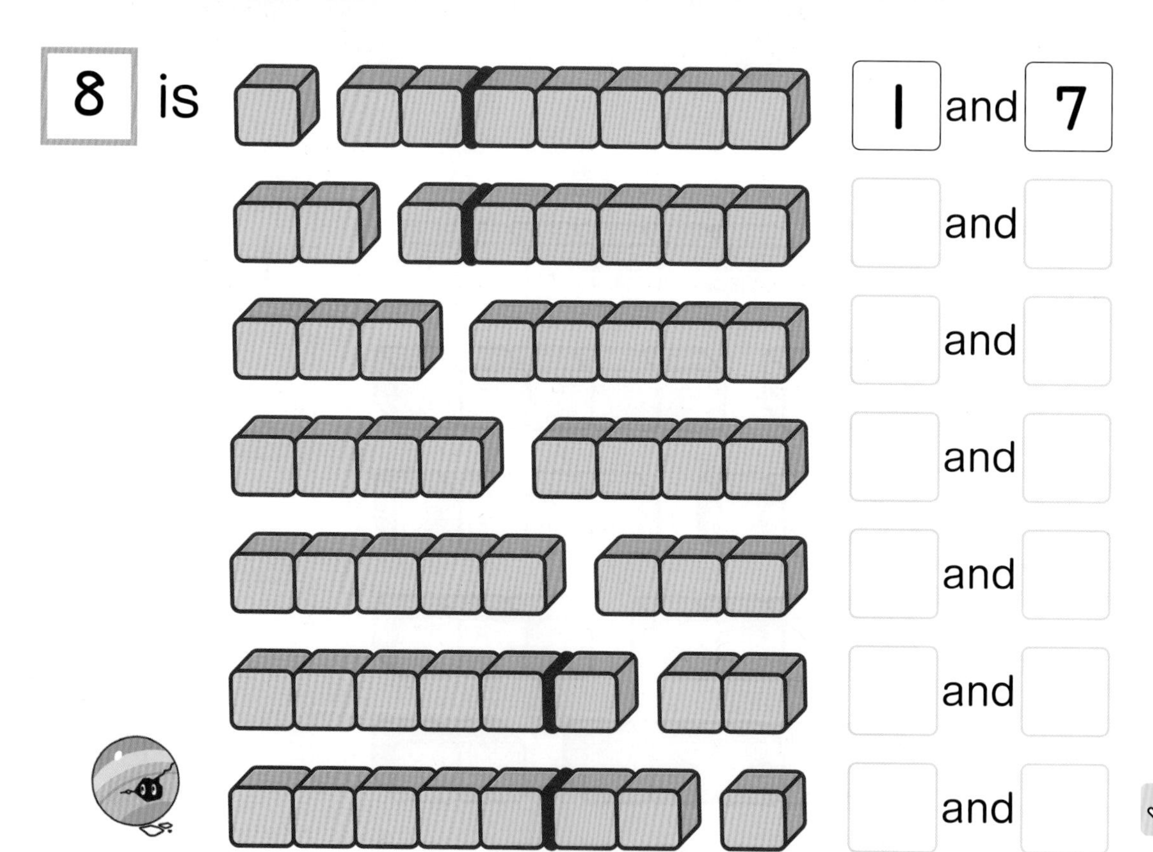

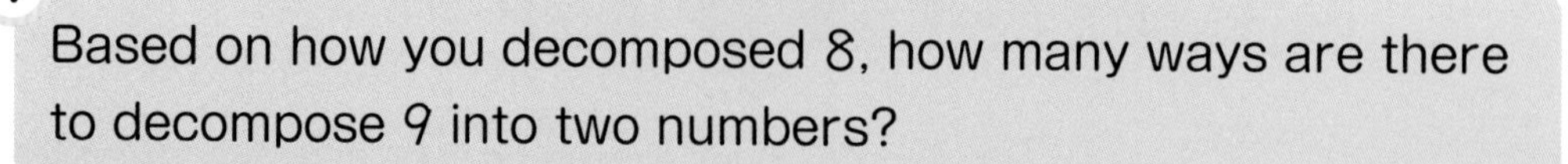

Based on how you decomposed 8, how many ways are there to decompose 9 into two numbers?

9

Let's make 9 by using two cards.

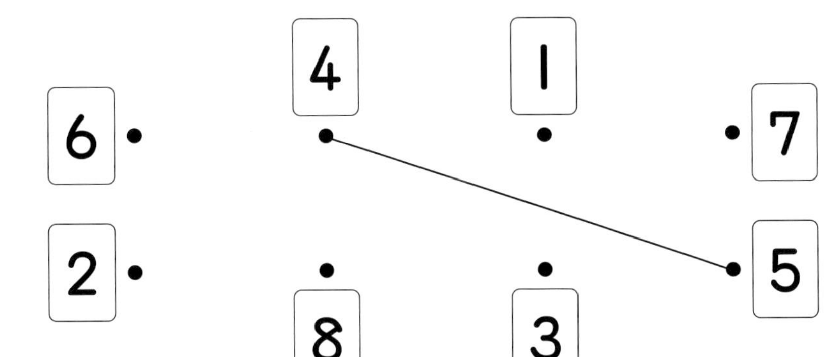

9	
4	5

9	
6	

9	

9	

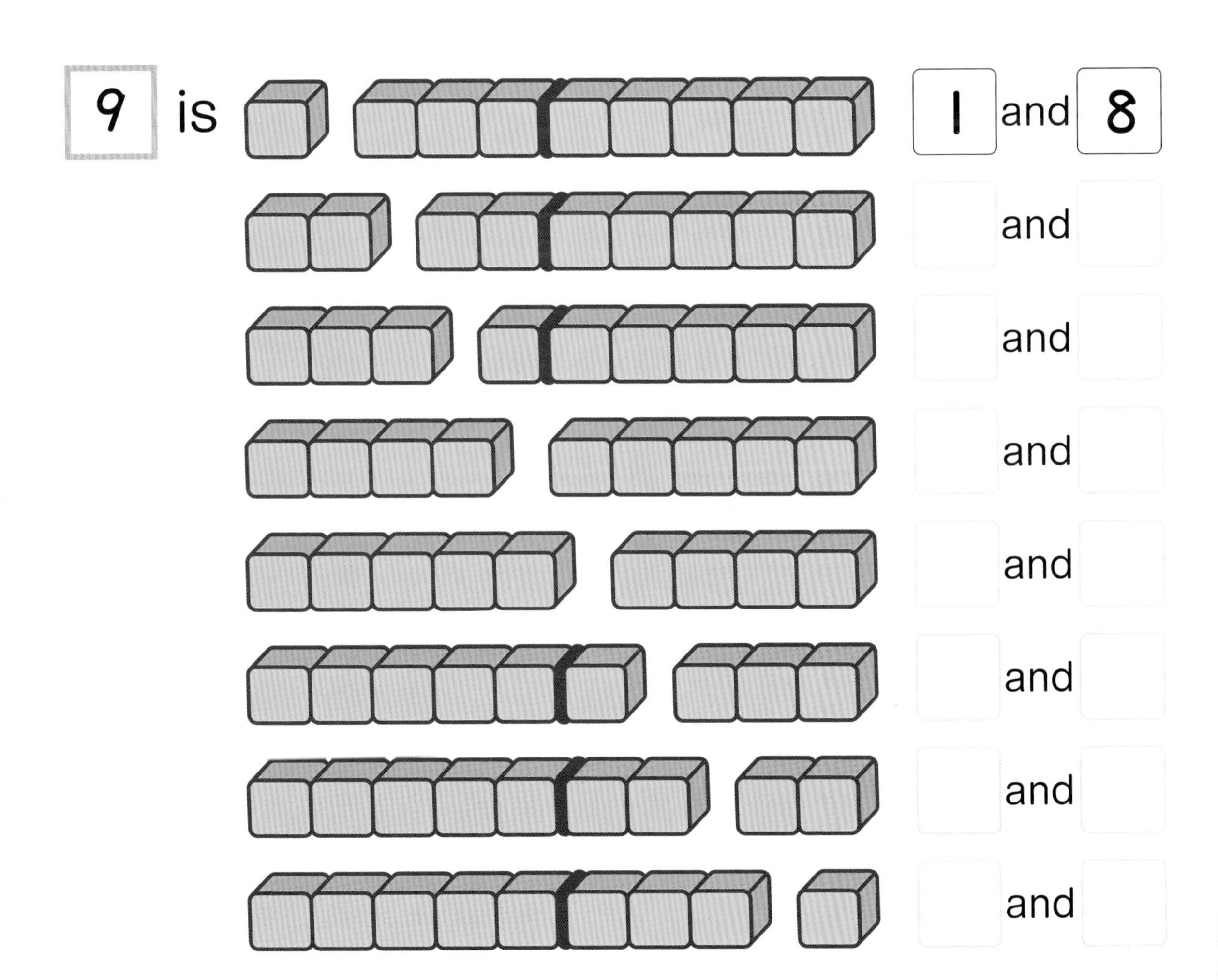

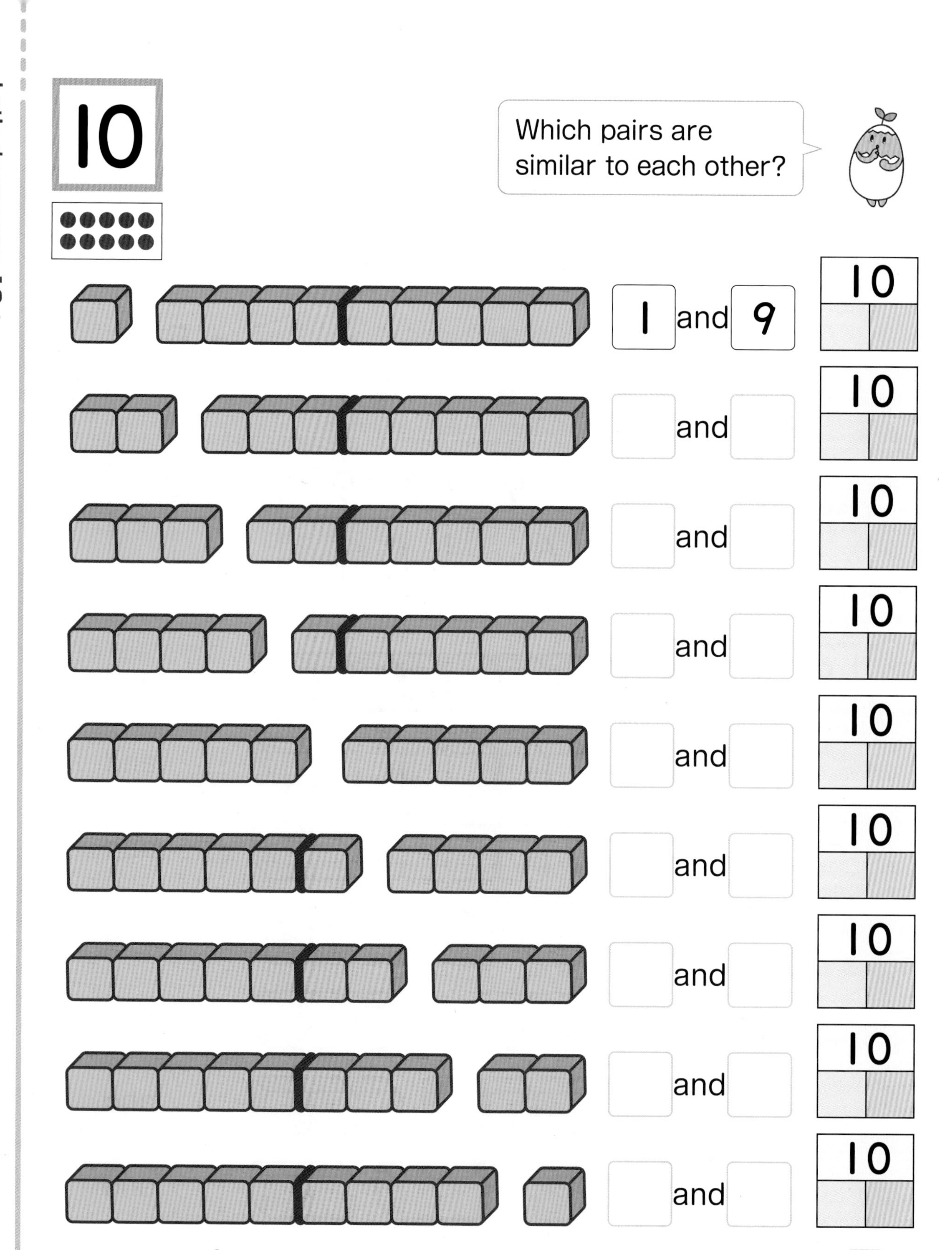

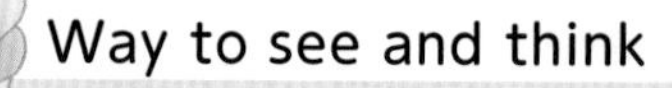

If you use the idea of decomposing, you can make so many pairs.

1 If you make 10, show your cards.

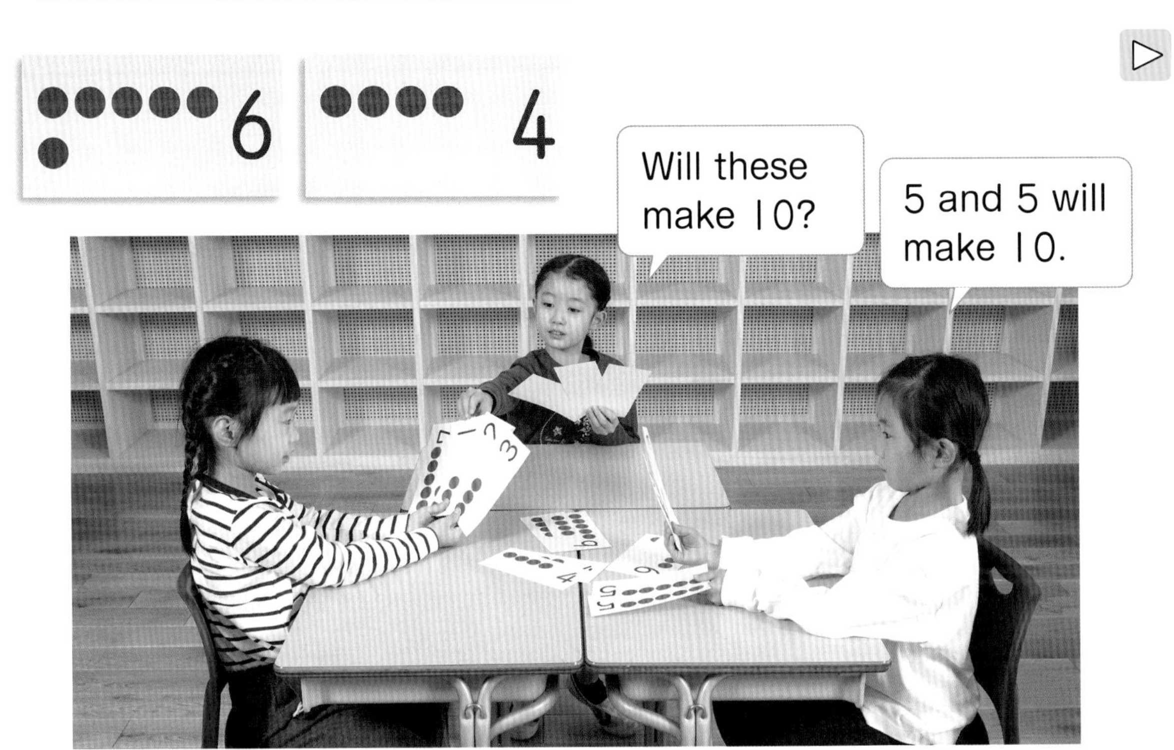

Let's circle two numbers that make 10.

5	8	2	6
5	1	3	4
6	9	7	5
4	2	8	5

2	6	4	3
1	8	7	8
9	6	5	2
4	7	3	5

Supplementary Problems → p.90

3 Ordinal Numbers

1 Look at the picture above and answer. ▷

① Where is the dog running from the front?

② What is the position of the panda?

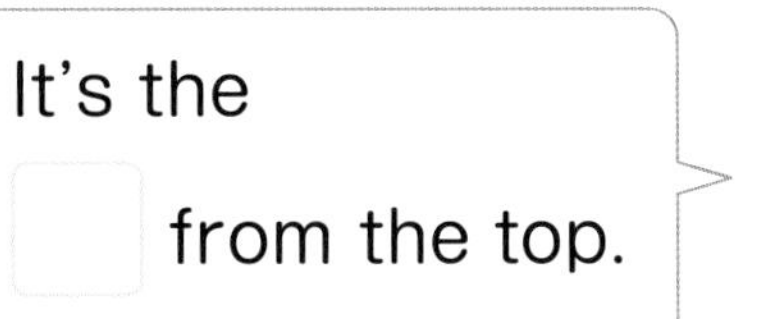

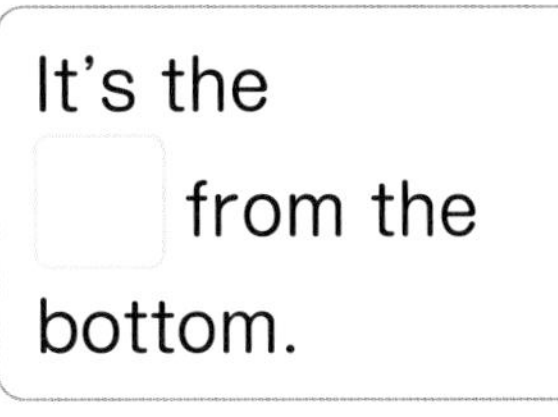

③ Let's do the same with the other animals.

The first 5 children can ride on the roller coaster. Kai is the 5th child from the front.

The first 5 children

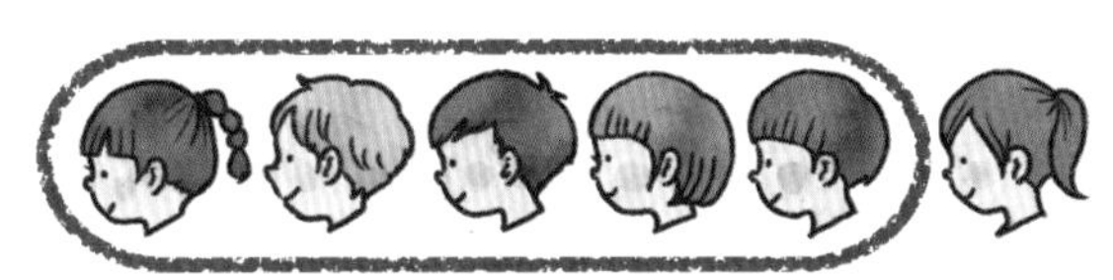

The 5th child

1 Let's color it!

① The first 2 cars.

② The 2nd car.

③ The 3rd car from the end of line.

2 Raise your hand when your position is called.

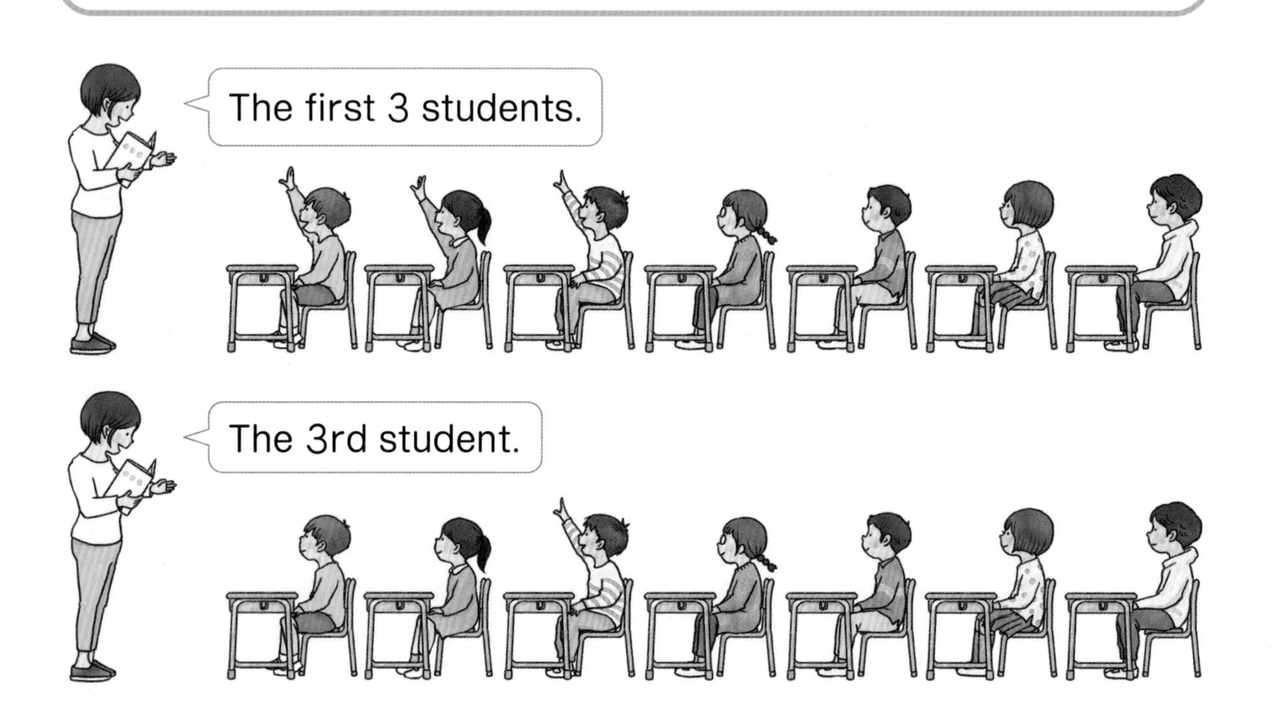

1 Guess where the same picture card is.

2 Explain where your classmates sit.

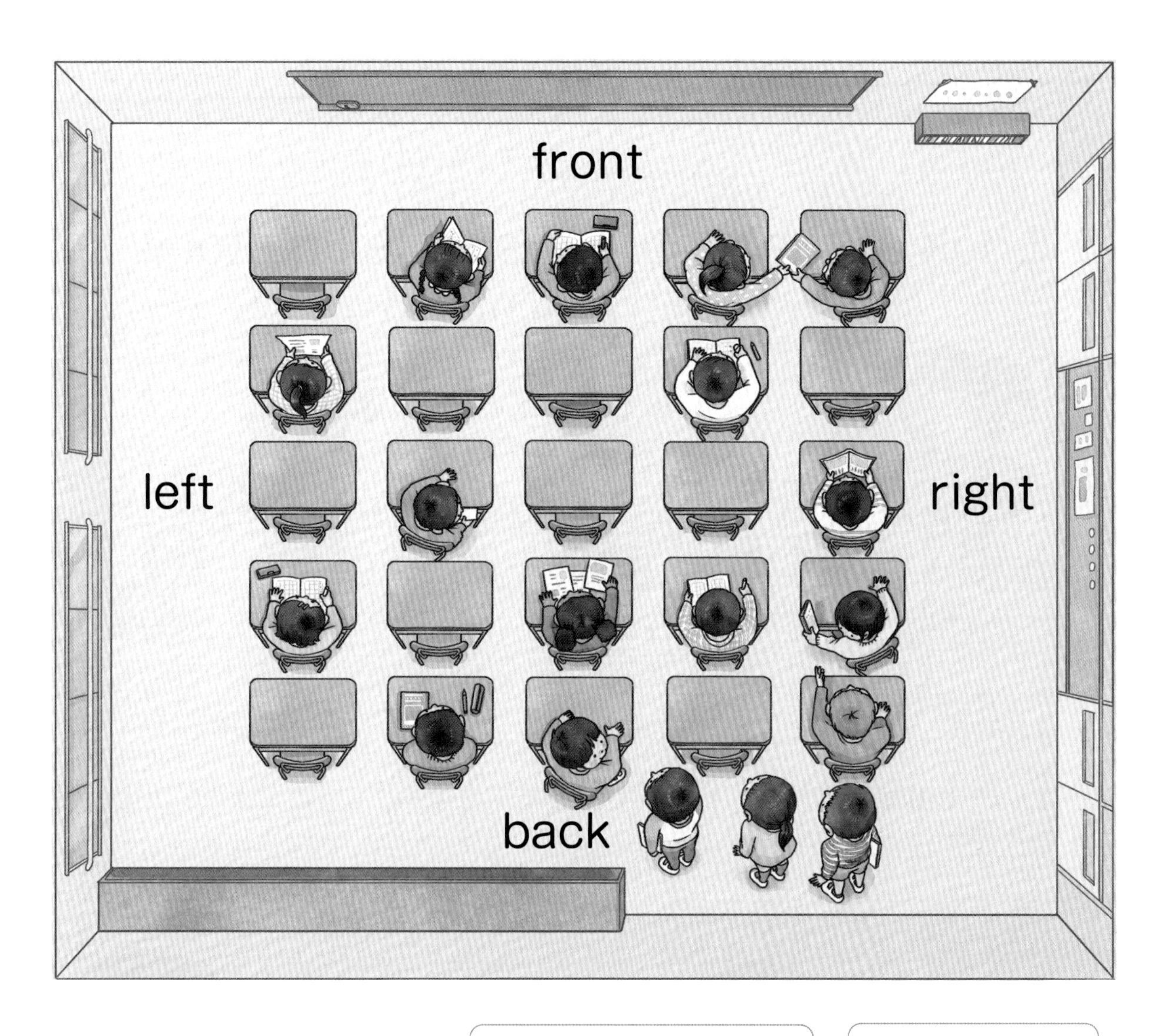

Kazuki's seat is the 3rd from the left, 2nd from the front.

Sakura's seat is the 2nd from the right, 3rd from the back.

Kota's seat is in the middle of the classroom.

Sara

Akari

Haruto

Supplementary Problems → p.91 Let's deepen. → p.97

How Many: Altogether and Increase

Look at the pictures and tell a story.

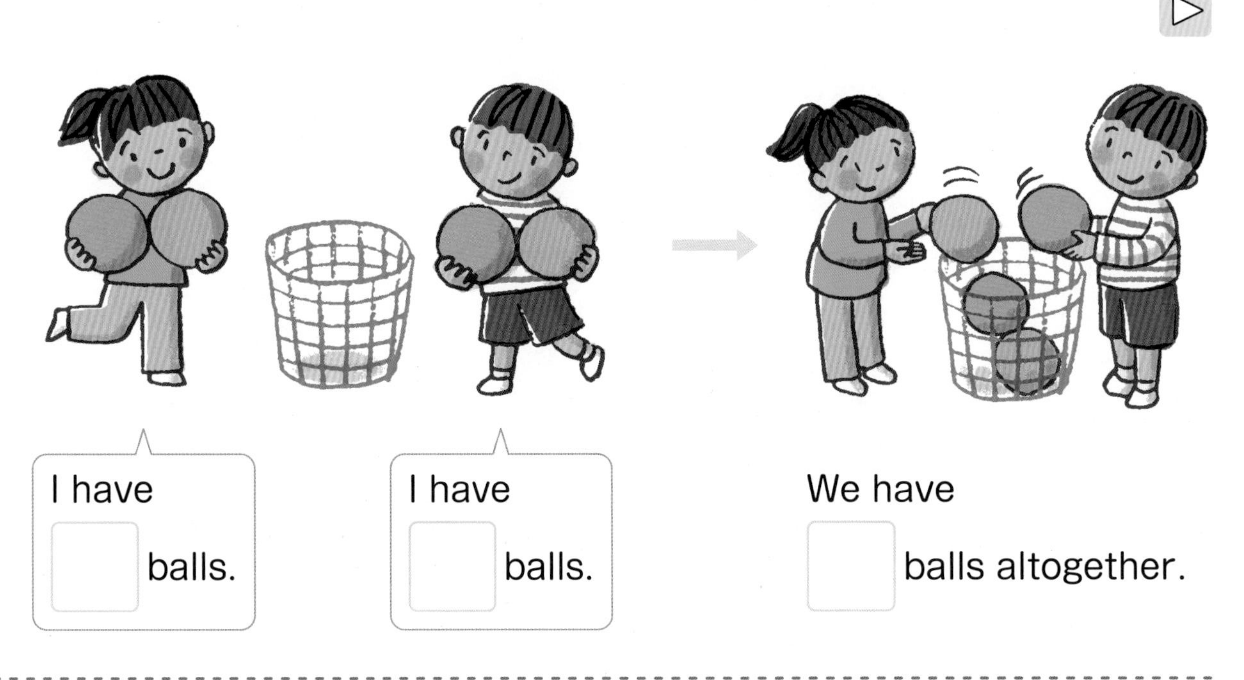

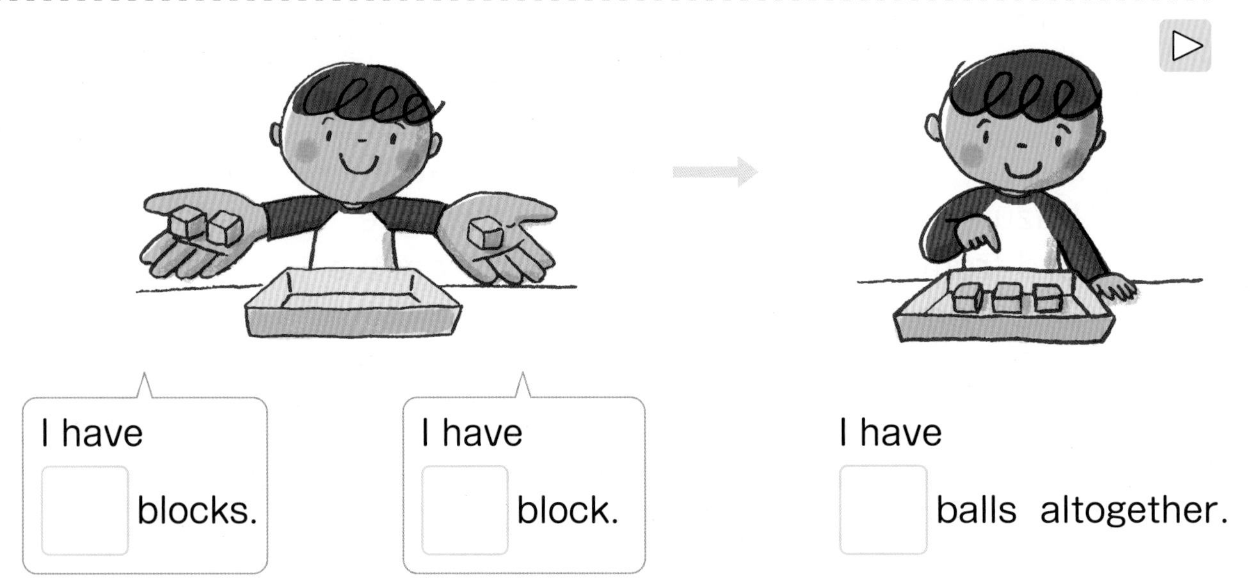

How many goldfish are there altogether?

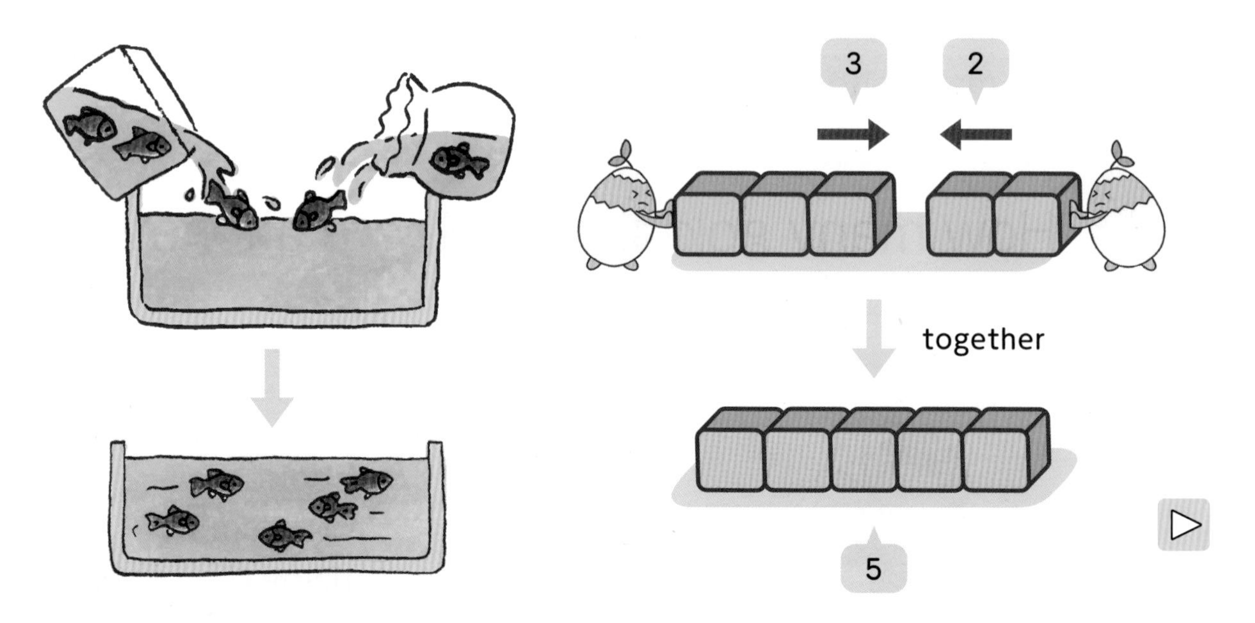

Putting 3 and 2 together makes 5.

Math Sentence:

$3 + 2 = 5$

3 plus 2 equals 5

Answer: 5 goldfish

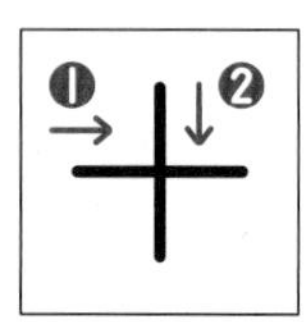

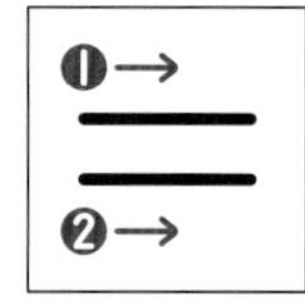

3+2=5 is a math sentence.
3+2 is a math expression. Math sentence and math expression can express our stories easily.

1 How many frogs are there altogether?

Math Sentence:

2 + 1 = ☐

Answer: ☐ frogs

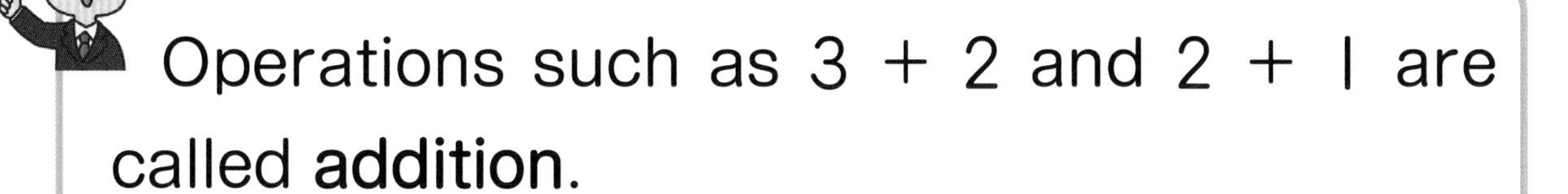

Operations such as 3 + 2 and 2 + 1 are called **addition**.

2 Let's do addition.

2 + 2	1 + 4	3 + 1	1 + 2
4 + 1	2 + 3	1 + 1	1 + 3

2

There are 5 red flowers and 4 white flowers. How many flowers are there altogether?

1 Which picture shows the situation, Ⓐ or Ⓑ?

② Let's write a math sentence and find out the answer.

Math Sentence: [ ]

Answer: [ ] flowers

③ Let's explain why you made the math sentence above using blocks.

1 Let's do addition.

$5 + 2$ $5 + 3$ $5 + 1$

Let's draw a picture →

3

There are 2 black cats and 5 white cats. How many cats are there altogether?

① Let's draw a picture of this situation.

Akari's picture

Yu's picture

Which one is quicker to draw?

Haruto

What is the good point in Yu's picture?

Sara

2 Let's write a math sentence in your notebook and find out the answer.

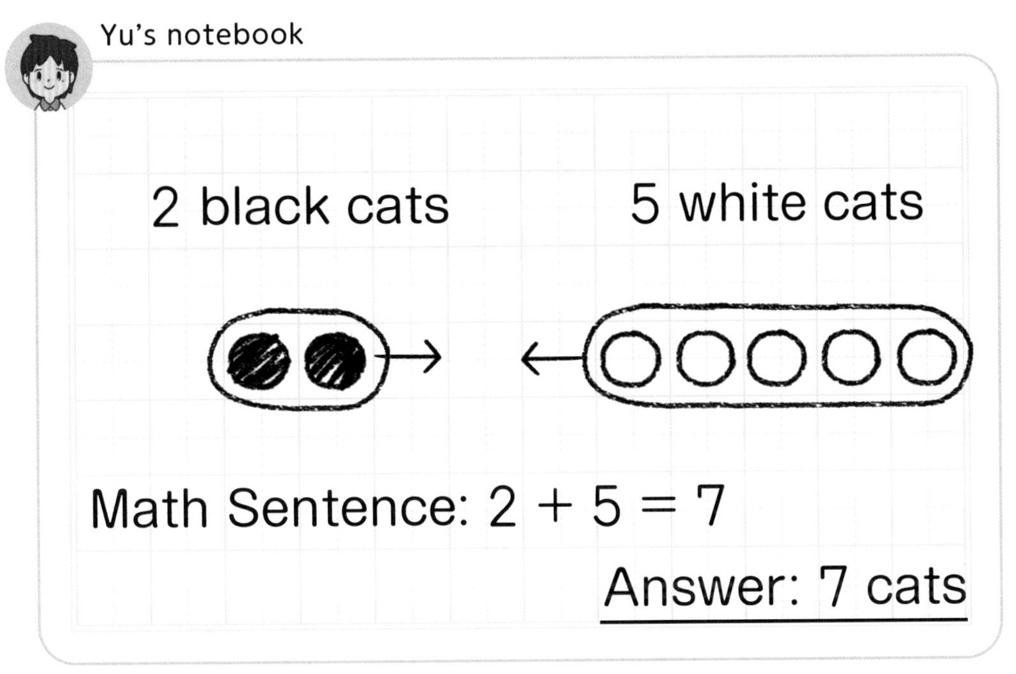

Way to see and think

It is clear to think of the problem using blocks, pictures and math sentences!

1 There are 3 white flowers and 5 red flowers. How many flowers are there altogether?

Let's draw a picture of the situation!

Math Sentence:

Answer: ☐ flowers

2 Let's do addition.

4 + 5　　　3 + 5　　　1 + 5

Look at the picture and tell a story.

How many increase? →

2 How many increase?

1 How many goldfish are there after you put more?

Let's think about the problem by using [block].

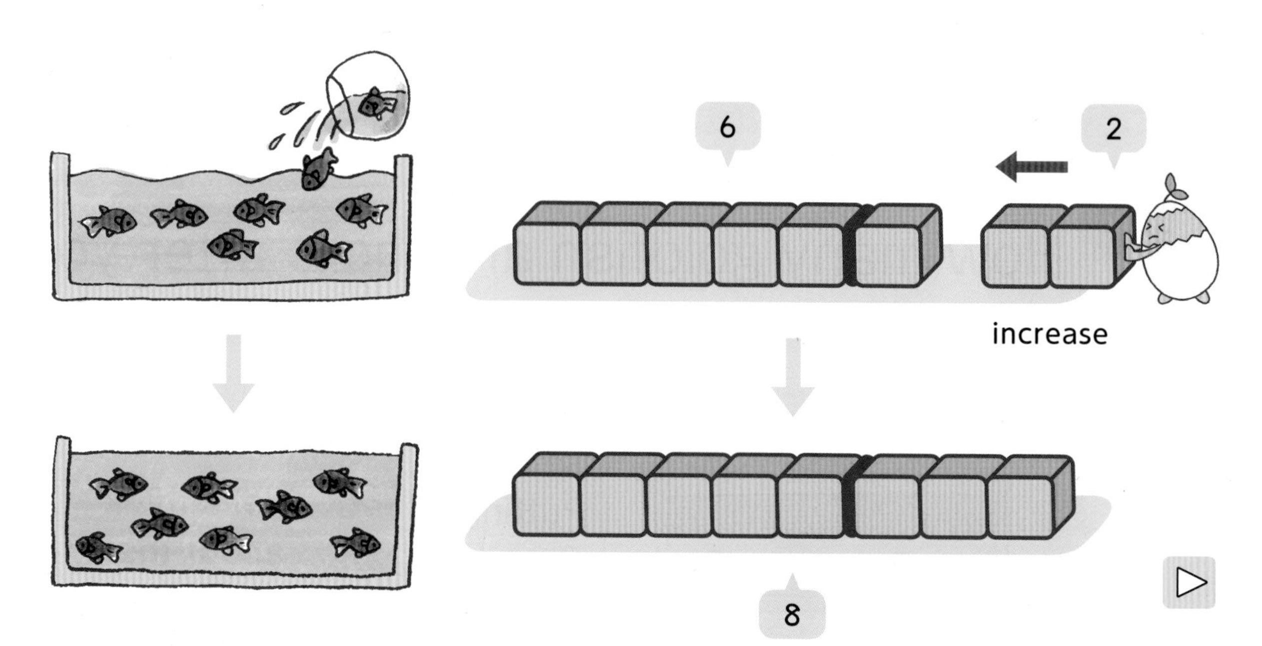

There are 6 goldfish. Put 2 more goldfish, and it becomes 8 goldfish.

Math Sentence:

$6 + 2 = 8$

Answer: ☐ goldfish

Sara: Same as "how many altogether", we use addition.

1 How many blocks are there altogether?

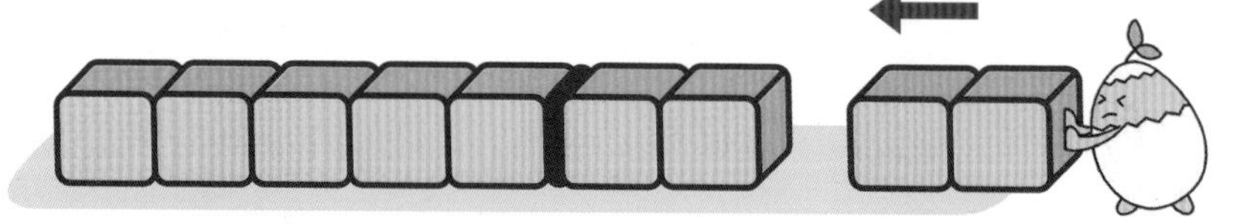

Math Sentence:

☐ + ☐ = ☐

Answer: ☐ blocks

2 Let's do addition.

$8+1$　　$6+3$　　$1+6$　　$2+6$

Let's make a math sentence →

2

There are 4 cars at the parking lot. When 3 more cars come into the parking lot, how many cars are there altogether?

1 Which picture shows the situation, Ⓐ or Ⓑ?

Ⓐ

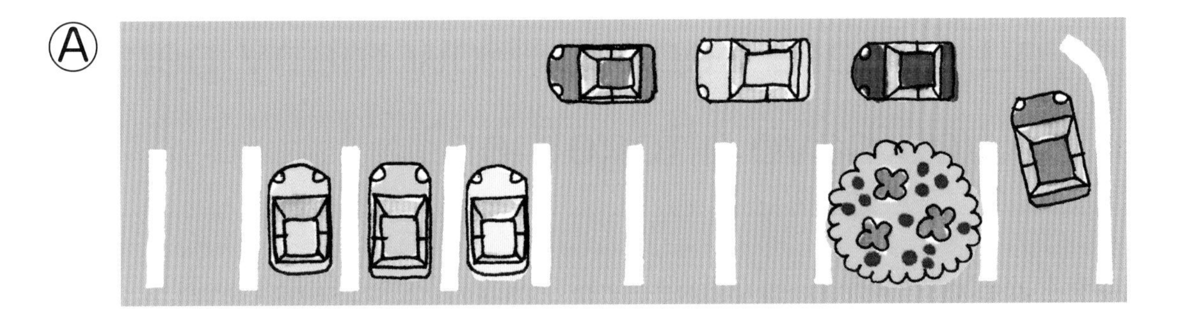

Ⓑ

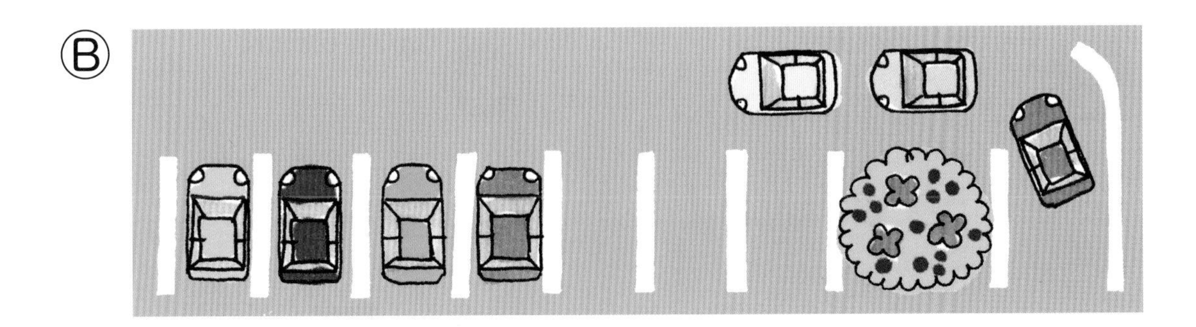

② Let's write a math sentence and find out the answer.

Math Sentence:

☐ + ☐ = ☐ Answer: ☐ cars

③ Let's explain why you made the math sentence above using blocks.

1 Let's do addition.

$2 + 4$ $3 + 3$ $3 + 4$ $4 + 4$

Let's draw a picture →

3

You have 6 pencils. Your sister gives you 4 more pencils. How many pencils do you have altogether?

① Let's draw a picture of this situation.

Haruto's picture

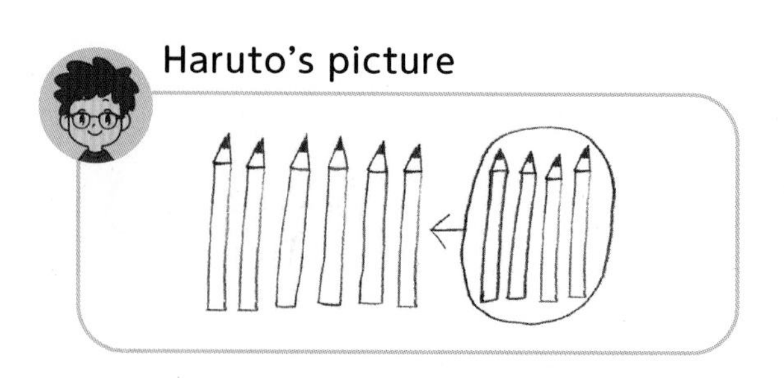

Sara's picture

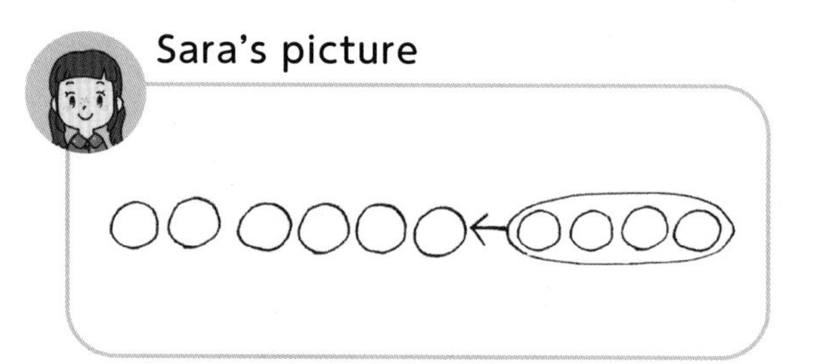

What is the good point in Sara's picture?

Yu

② Let's write a math sentence in your notebook and find out the answer.

Sara's notebook

At first I had 6 pencils.

My sister gives me 4 pencils.

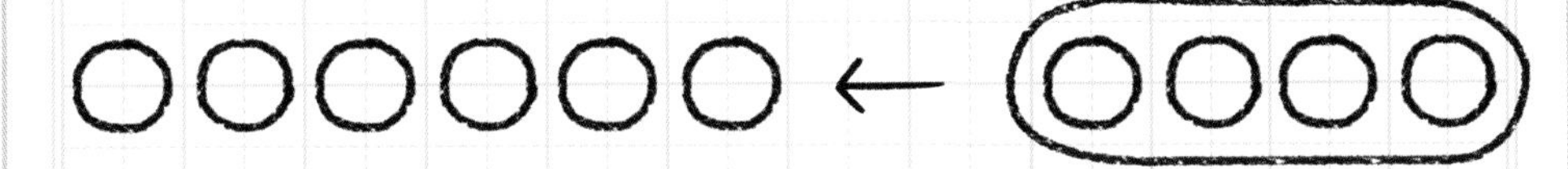

Math Sentence: $6 + 4 = 10$

Answer: 10 pencils

It is easier to write by using ○.

1 3 children are playing. 7 more children joins. How many children are there altogether?

Let's draw a picture.

Math Sentence:

Answer: ☐ children

2 Let's do addition.

$9 + 1$	$5 + 5$	$4 + 6$	$2 + 8$
$7 + 3$	$8 + 2$	$1 + 9$	$3 + 7$

4 Let's make a math story for 5 + 3.

① Look at the picture below, and make a math story.

There are ☐ monkeys.

There are ☐ monkeys.

How many monkeys are there ☐ ?

② Look at the picture below, and make a math story.

1 Let's make two math stories for 7 + 3.

Can you make a similar story like 4 ?

Haruto

Let's make a math story and share it with your friends.

2 Let's make a math story focusing on the movement of the blocks.

① 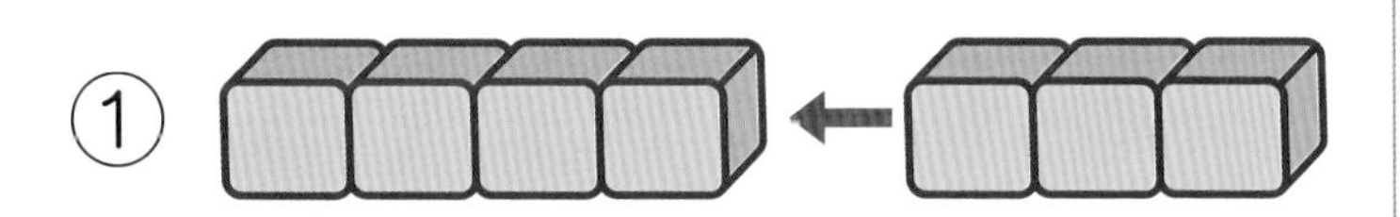

There are 4 blocks first, and 3 increased.

Akari

② 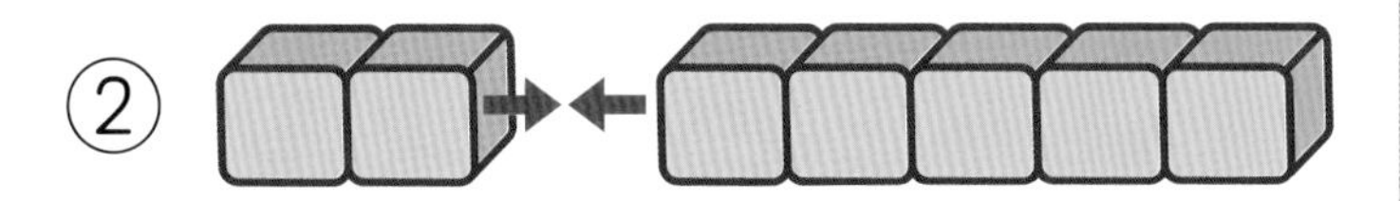

It is making 2 blocks and 5 blocks together.

Sara

Addition can be used in both situations.

Yu

Addition Cards

Let's make addition cards and practice addition facts.

card

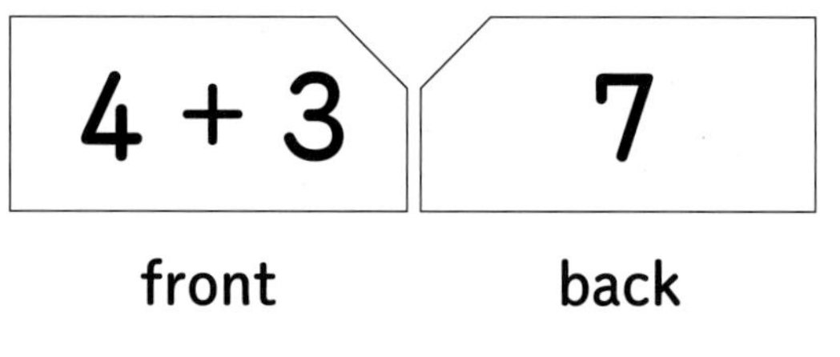

1 Say the answer.

2 Find other cards with the same answer.

3 Line up the cards with the same answer in order.

Decide your original rules and play with the cards!

3 Adding 0

Adding 0 →

1 You have two turns to throw balls into a basket. How many balls are there in the basket altogether?

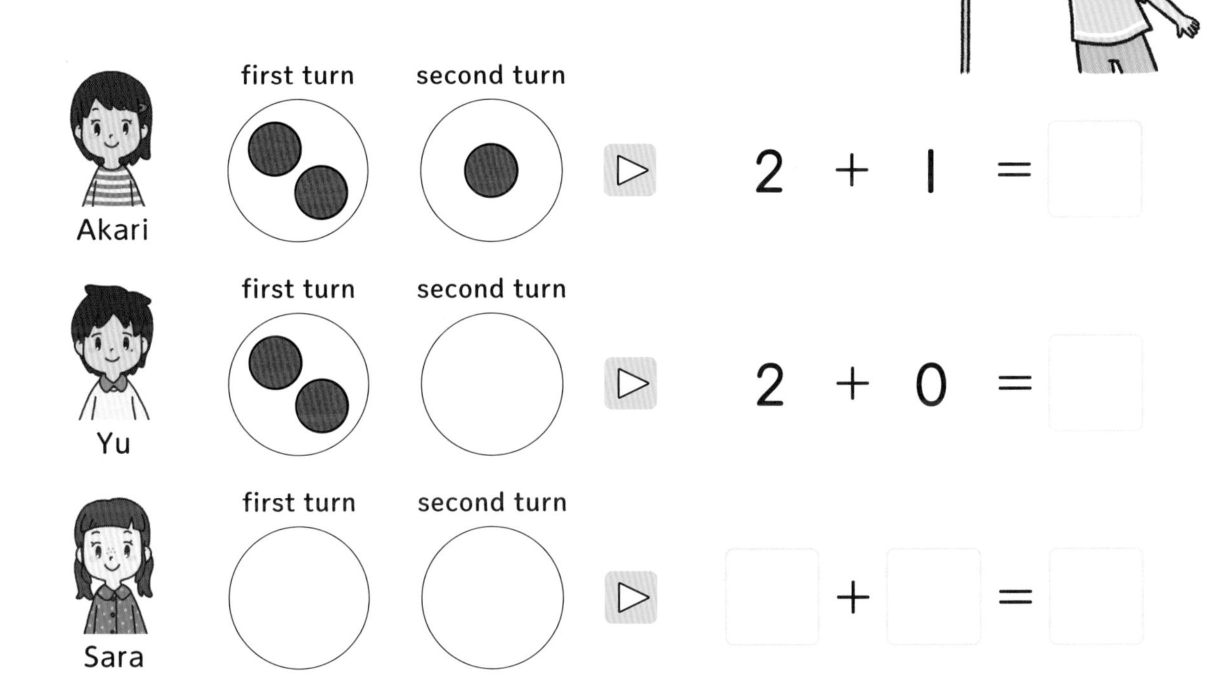

1 After throwing balls, 0 + 4 is written as the math expression. How many balls were inside the basket on each turn?

2 Let's do addition.

4 + 0	9 + 0	7 + 0	8 + 0
0 + 6	0 + 5	0 + 1	0 + 0

for Addition

Let's make a story for addition.

Together

There are 4 red cups.

There are 5 blue cups.

There are 9 cups altogether.

Increase

There are 3 beetles.

2 more beetles fly in.

There are 5 beetles altogether.

CAN What can you do?

☐ We can do addition. → pp.38 ~ 51

1 Let's do addition.

2 + 6	0 + 3	6 + 1	0 + 9
2 + 7	1 + 8	5 + 4	3 + 3
5 + 0	2 + 4	6 + 4	7 + 1
6 + 0	8 + 2	1 + 7	3 + 4

☐ We can find the addition expressions with the same answer. → pp.38 ~ 51

2 Let's connect the cards with the same answers.

3 + 5 •	• 4 + 4
2 + 4 •	• 6 + 3
4 + 5 •	• 5 + 1

☐ We can make an addition expression and find the answer. → pp. 43 ~ 47

3 There are 6 candies.
You get 2 more candies.
How many candies do you have altogether?

Supplementary Problems → p.92 Let's deepen. → p.98

How Many: Left and Difference

Look at the picture below, and explain the situation.

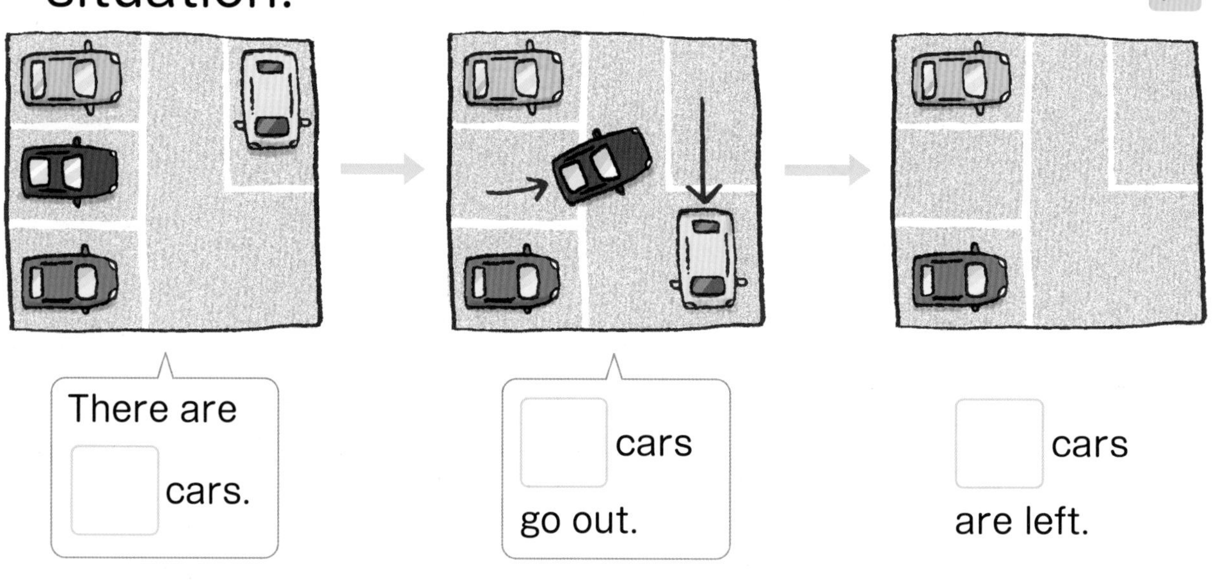

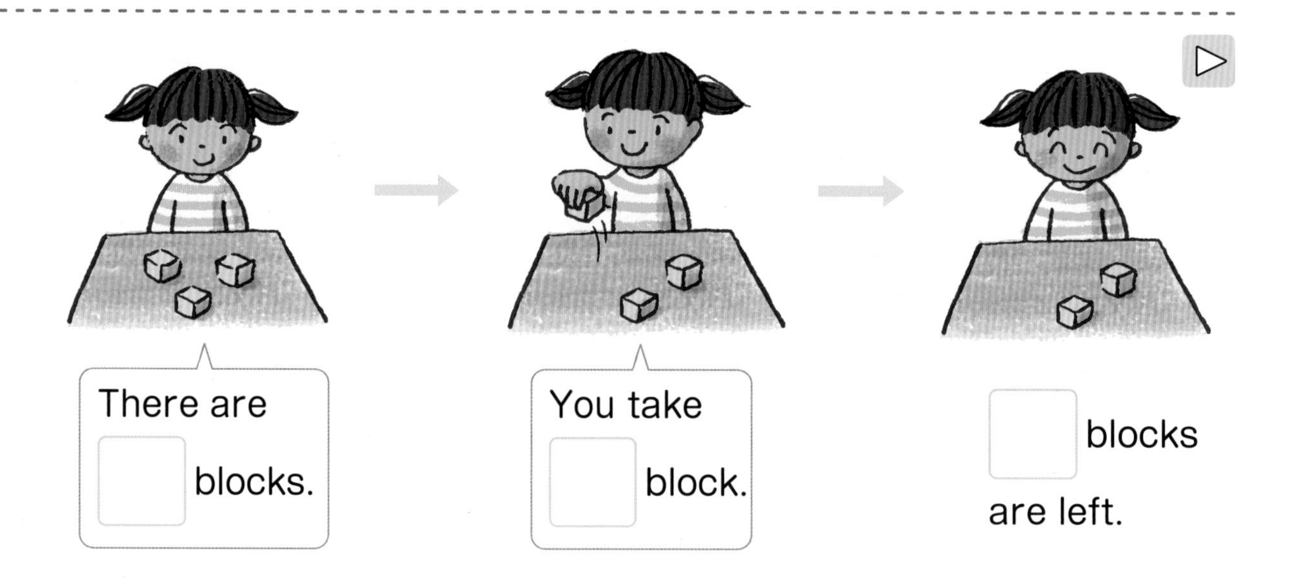

How many goldfish are left?

Let's think about the problem by using [block].

Way to see and think

It's easier to understand by using blocks!

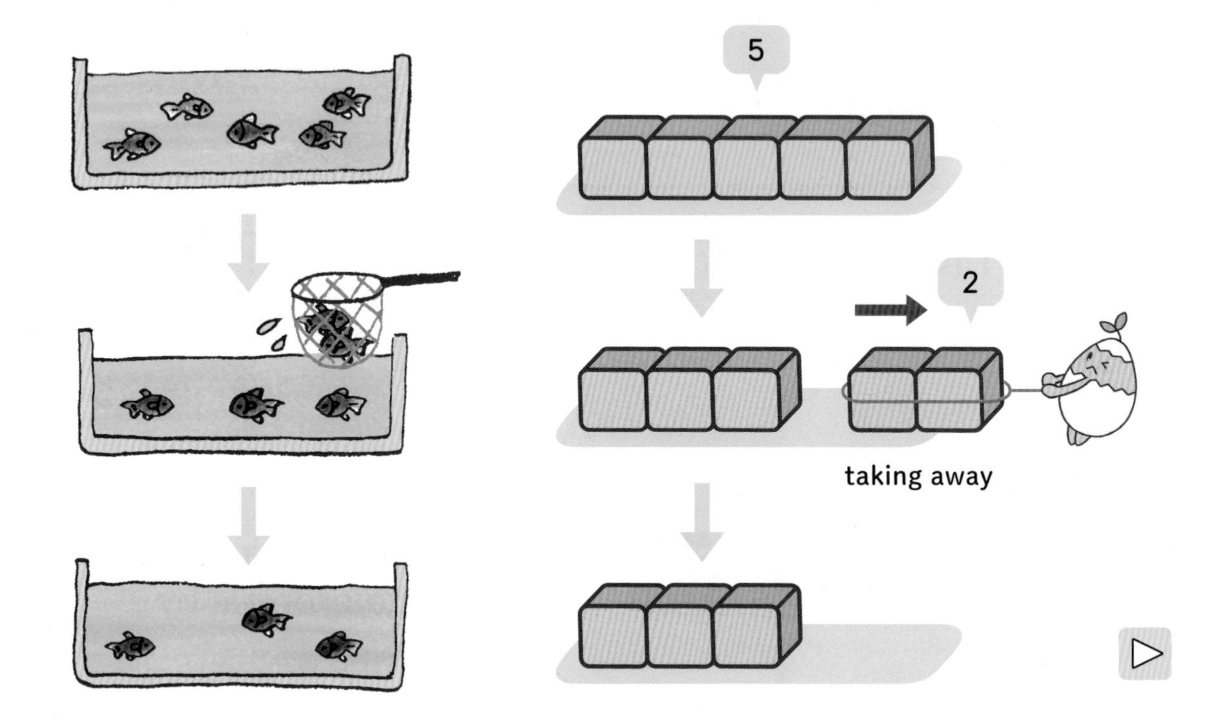

Taking away 2 from 5 is 3.

Math Sentence:

5	−	2	=	3
5	minus	2	equals	3

Answer: 3 goldfish

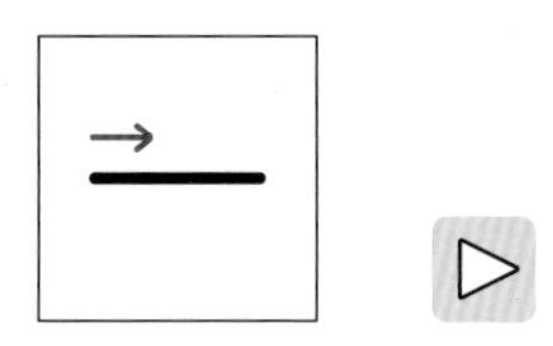

5−2=3 is a math sentence. 5−2 is a math expression.

1 How many are left?

Math Sentence:

□ − □ = □

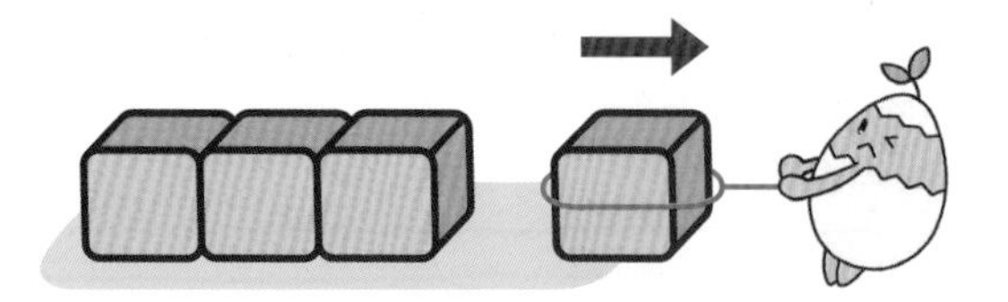

Answer: □ pieces

Operations such as 5 − 2 and 4 − 1 are called **subtraction**.

2 Let's do subtraction.

5 − 3	2 − 1	4 − 2	5 − 4
4 − 3	3 − 1	5 − 1	5 − 2

Let's think of a subtaction math sentence →

2

There were 9 sheets of origami paper. You used 4 sheets to make paper airplanes. How many sheets were left?

1 Which picture shows the situation, Ⓐ or Ⓑ?

Ⓐ
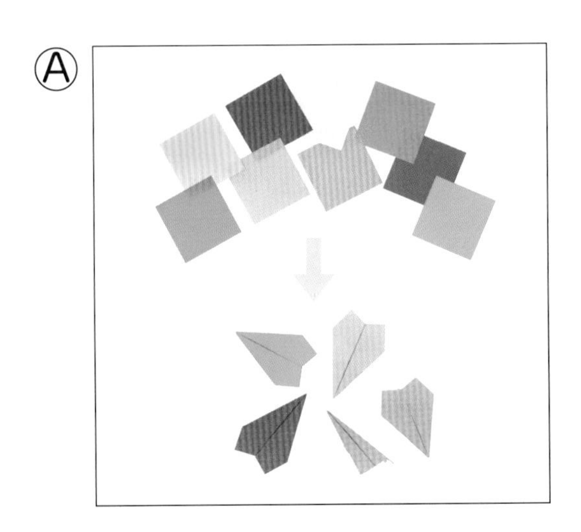

Ⓑ
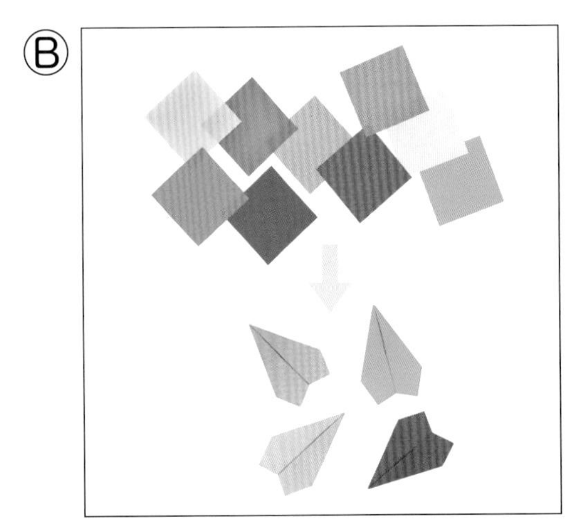

② Let's write a math sentence and find out the answer.

Math Sentence:

Answer: ☐ sheets

③ Let's explain why you made the math sentence above using blocks.

1 Let's do subtraction.

7 − 2　　8 − 3　　6 − 5　　8 − 5

3

There were 9 children playing. After a while, 3 children went home. How many children are still playing?

① Let's draw a picture.

Akari's picture

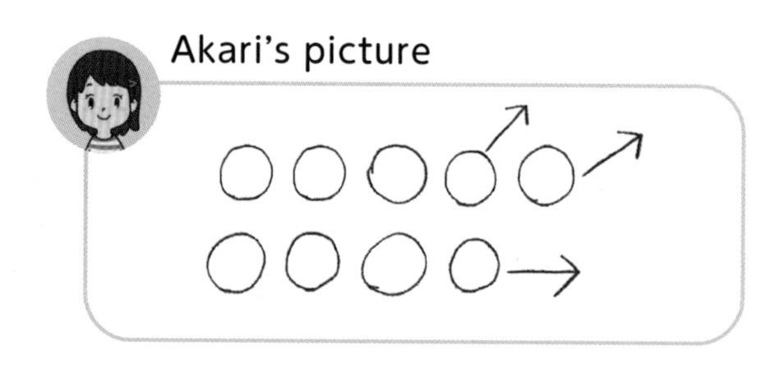

Yu's picture

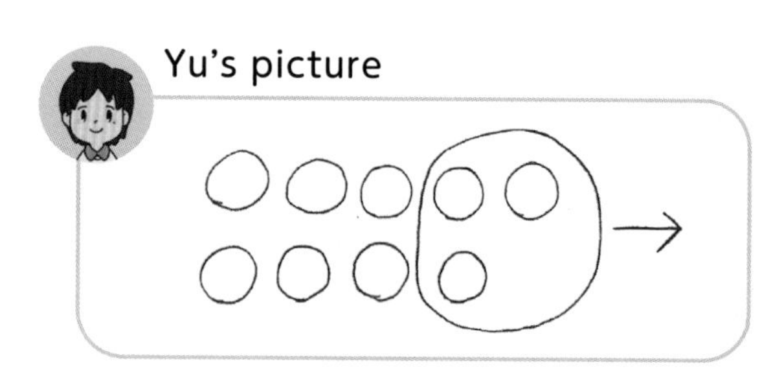

Even when do addition, we drew pictures and thought about them.

Haruto

Let's draw a picture →

② Let's write a math sentence in your notebook and find out the answer.

Yu's notebook

There were 9 children. 3 children went home.

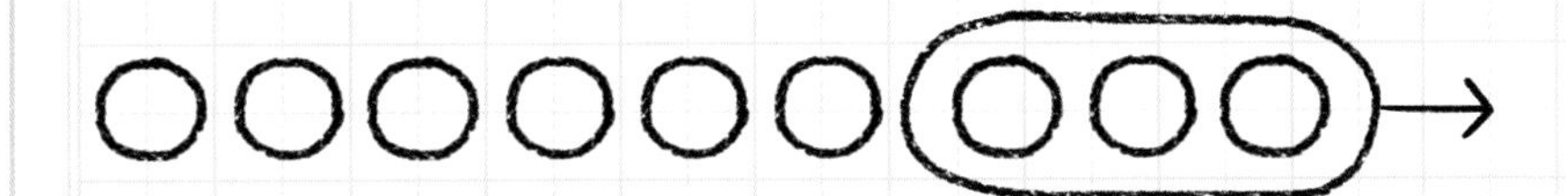

Math Sentence: $9 - 3 = 6$

Answer: 6 children

1 There were 8 flowers. I gave 4 of them to my friend. How many flowers are left?

Math Sentence:

Answer: ☐ flowers

2 Let's do subtraction.

$9 - 1$	$8 - 2$	$7 - 1$	$9 - 2$
$9 - 7$	$7 - 6$	$6 - 3$	$8 - 7$
$6 - 4$	$7 - 3$	$9 - 6$	$7 - 4$

4

There are 10 dogs. There are 6 .
How many are there?

(1) Let's discuss how to make a math sentence for this situation with your friends.

(2) Let's write a math sentence and find out the answer.

Math Sentence: ________ Answer: ____ dogs

1 There were 10 pencils. He sharpened 3 of them. How many pencils are unsharpened?

2 Let's do subtraction.

$10-4$	$10-1$	$10-9$	$10-2$
$10-6$	$10-8$	$10-7$	$10-5$

2 Subtracting 0

1 You are taking away goldfish from the water tank. How many goldfish are left?

Taking away 1. ▷ 3 − 1 = ☐

Taking away 2. ▷ 3 − ☐ = ☐

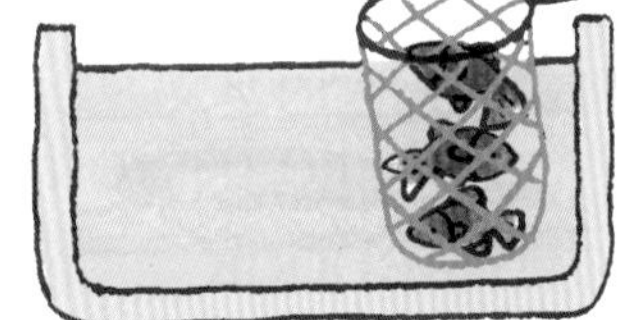

Taking away 3. ▷ 3 − ☐ = ☐

Taking away 0. ▷ 3 − 0 = ☐

1 Let's do subtraction.

7 − 7　　4 − 4　　5 − 5　　9 − 9

8 − 0　　1 − 0　　6 − 0　　0 − 0

1 How many more cows are there than horses?

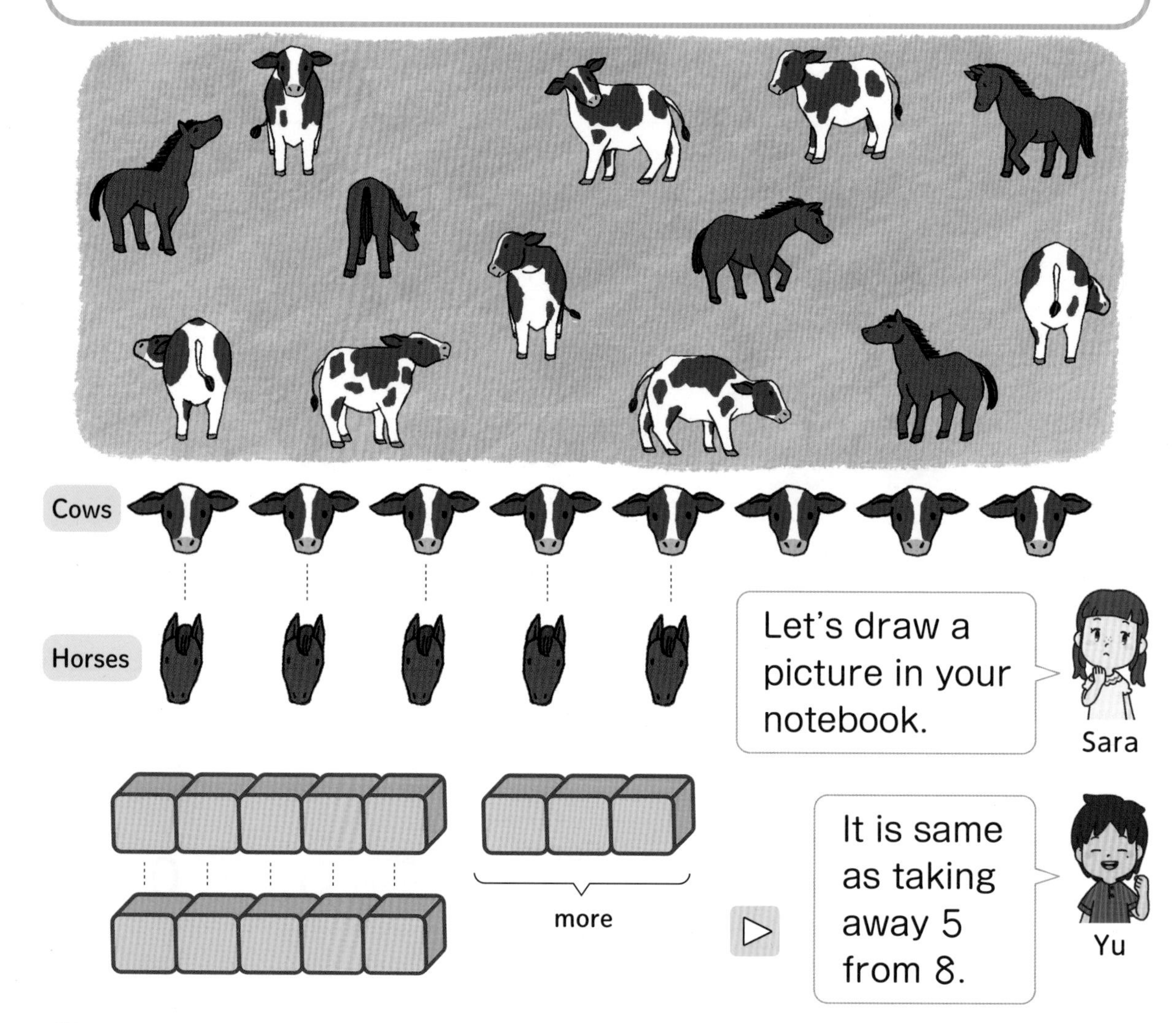

① Let's write a math sentence and find out the answer.

Math Sentence:

$8 - 5 = \square$ Answer: $\square$ more cows

1 How many fewer trucks are there than buses?

Math Sentence: ☐ − ☐ = ☐

Answer: ☐ fewer trucks

2 There are cats and dogs. Which is fewer and by how many?

Math Sentence: ☐ − ☐ = ☐

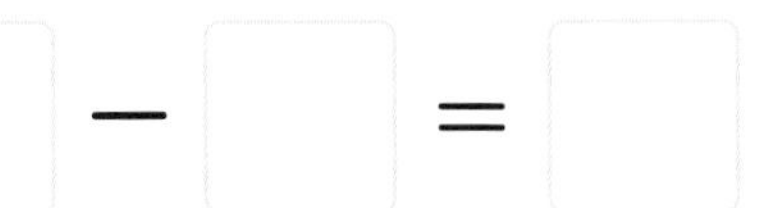

Answer: There are ☐ fewer ☐ .

2 Look at the picture below, and let's make a math story for $8 - 6$.

❶

There were ☐ swallows on the wire.
☐ swallows flew away. How many swallows are ☐ ?

❷

There are ☐ yellow cars.
There are ☐ red cars.
How many more ☐ cars are there than ☐ cars?

1 Let's make two math stories for $9 - 4$.

Akari

Can you make similar questions like 2 ?

Let's make math stories and share it with your friends.

2 Let's make a math story focusing on the movement of the blocks.

① 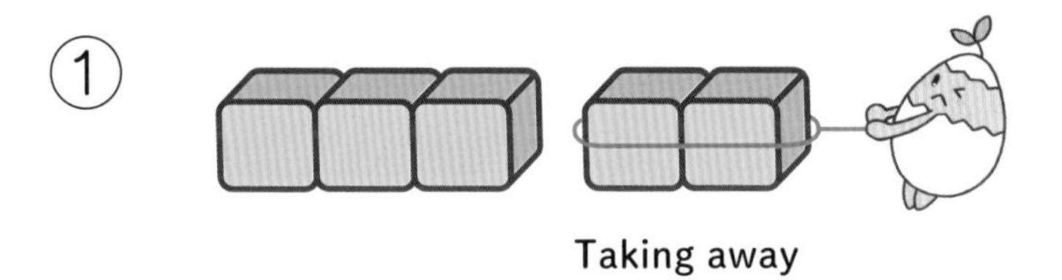

He is trying to take away 2 blocks from 5 blocks.

Sara

② 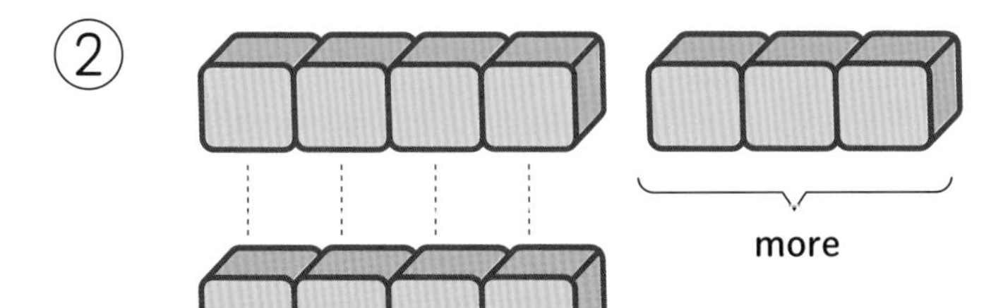

It is focusing on the difference between 7 blocks and 4 blocks.

Yu

Subtraction can be used in both situations.

Haruto

3 Each child gets 1 candy. How many candies will be left? ▷

Math Sentence: ☐ Answer: ☐ candies

1 You want to put 1 cake on each plate. How many more plates do you need?

Math Sentence: ☐ Answer: ☐ more plates

Subtraction Cards

Let's make subtraction cards and practice subtraction facts.

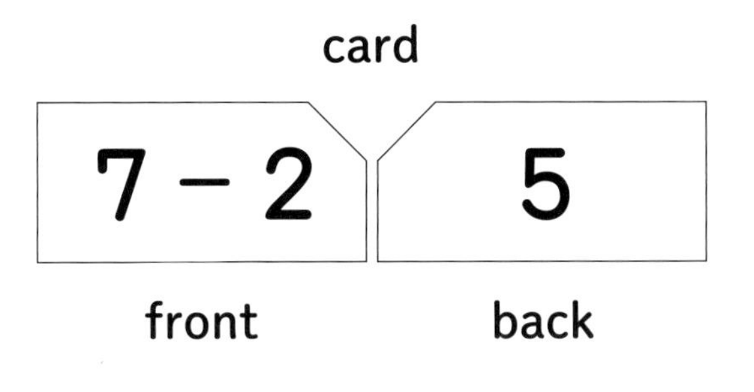

1 Say the answer.

2 Find other cards with the same answer.

3 Line up the cards with the same answer in order.

Decide your original rules and play with the cards!

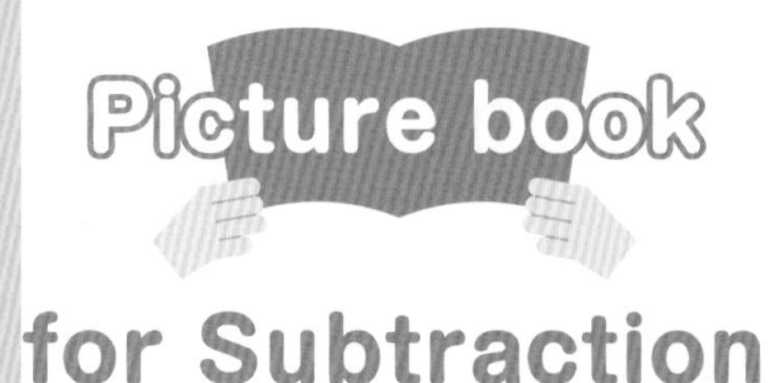

for Subtraction

Let's make a story for subtraction.

There were 6 bananas.

He ate 2 bananas.

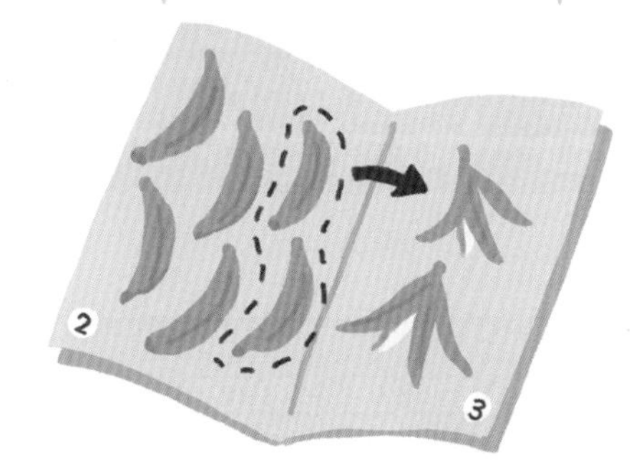

There are 4 bananas left.

There are 4 oranges.

There are 3 apples.

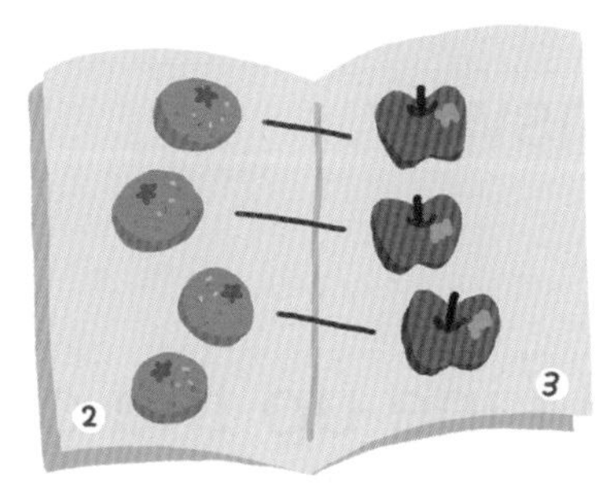

There is 1 more orange than apples.

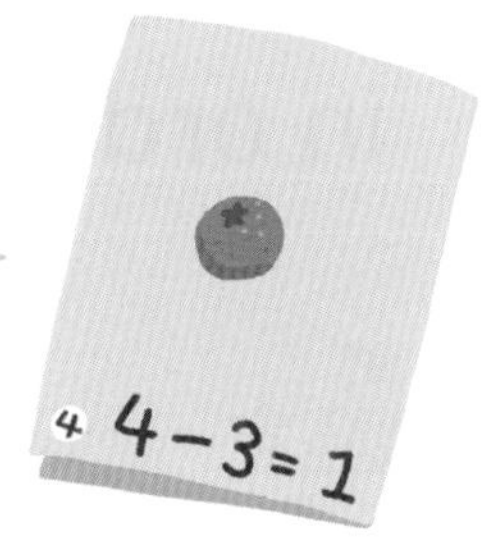

CAN What can you do?

☐ We can do subtraction. → pp.56 ~ 61

1 Let's do subtraction.

$4-1$	$9-8$	$2-2$
$6-2$	$7-5$	$8-8$
$7-0$	$10-3$	$8-1$

☐ We can make a subtraction expression and find out the answer. → pp.58 ~ 63

2 Let's write a math expression and find out the answer.

① There were 8 apples. You ate 4 apples. How many apples are left?

② There are 6 pencils and 10 crayons. Which is more and by how many?

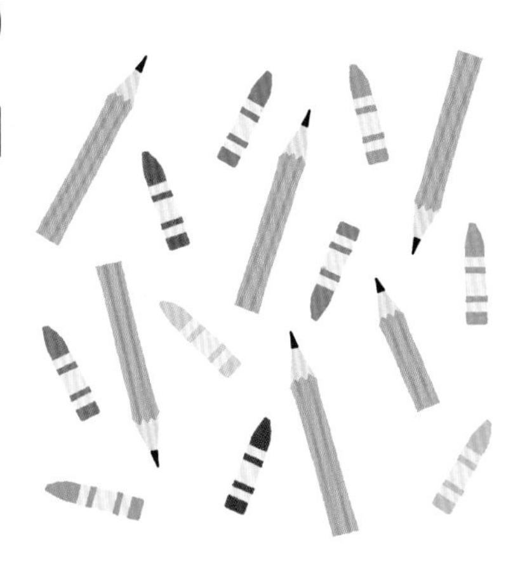

Supplementary Problems → p.93

Utilizing Math for SDGs

Story Making Activity for Addition and Subtraction

You will do group work in math class. Everyone looks like they are having fun.
Look around your classroom, and let's create a story using addition and subtraction expression!

① Let's try to make stories and expressions of addition and subtraction with the tools on the desks shown on the right page.

There were 5 pencils on the desk of Group 3. 2 pencils were handed to Mana in Group 2. How many pencils are left on the desk of Group 3?

3 pencils were left.

② Let's make a presentation of your story.

First of all, the order of the presentation should be decided. What is your presentation turn? Now, let's give a presentation in order.

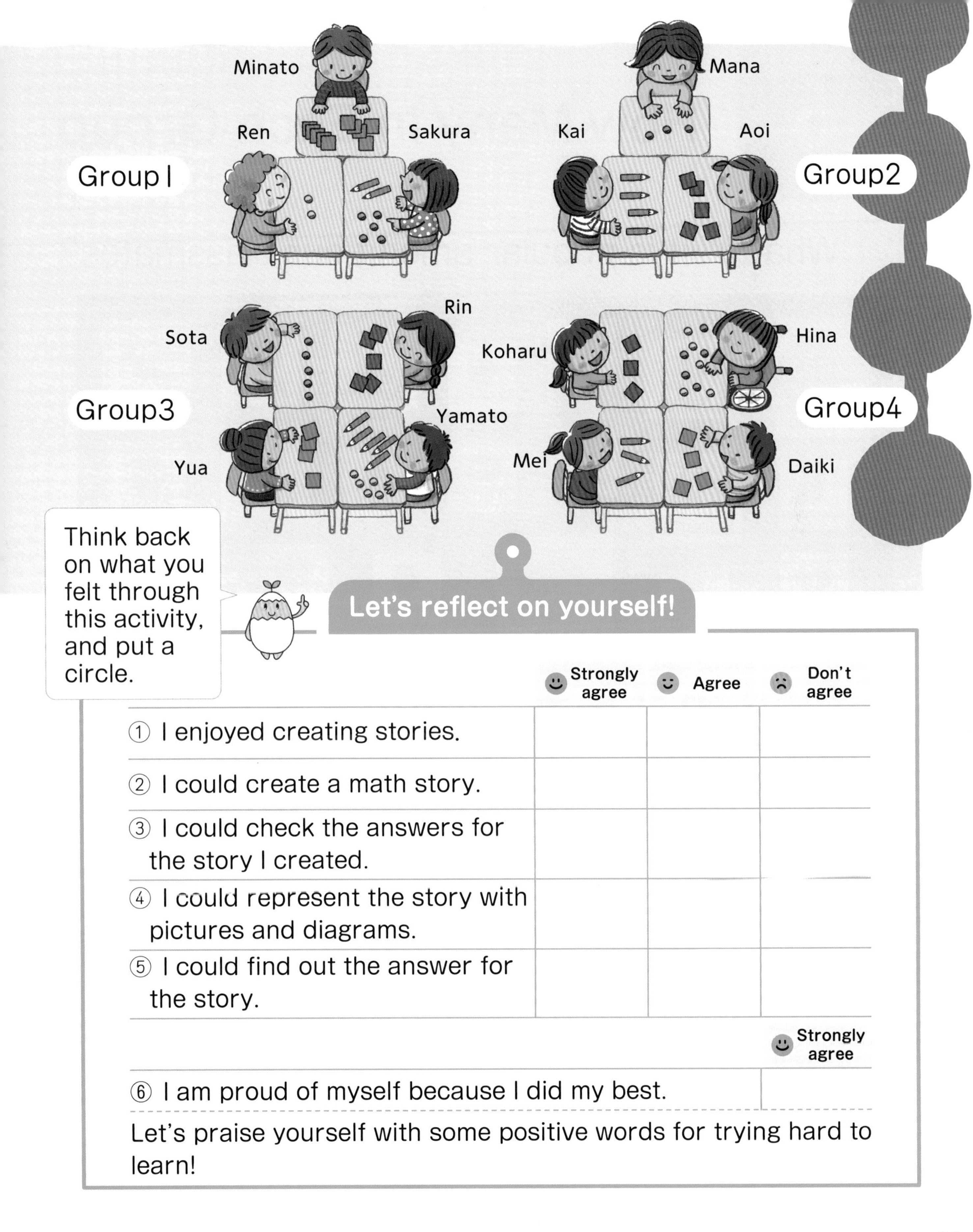

Let's reflect on yourself!

	Strongly agree	Agree	Don't agree
① I enjoyed creating stories.			
② I could create a math story.			
③ I could check the answers for the story I created.			
④ I could represent the story with pictures and diagrams.			
⑤ I could find out the answer for the story.			

	Strongly agree
⑥ I am proud of myself because I did my best.	

Let's praise yourself with some positive words for trying hard to learn!

6 How Many in Each Group

What fruit is popular among our classmates?

Let's think using the cards on page 99.

1 Let's examine the number of fruits.

I'll try to line them up.

Sara

But they don't have the same size.

Yu

How shall I count them?

Akari

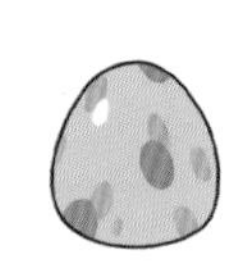

Examining the Fruits →

1. Let's think about the ways of counting with your friends.

2. Let's color the number of fruits in each group.

Let's color from the bottom one.

3. Which group of fruits has the largest group?

4. What is the fruit of which number is 6?

Let's tell your friend more about what you found out.

Way to see and think

It's easier to count if you make them the same size.

7 Numbers Larger than 10

1 Numbers up to 20

Can all the squirrels eat 1 acorn each?

Yu

How many flowers are there altogether?
Sara
There are 10 .
Akari

1 Let's count the number of squirrels.

10 and 3

13 squirrels

1 Let's count the number of acorns.

10 and 5

acorns

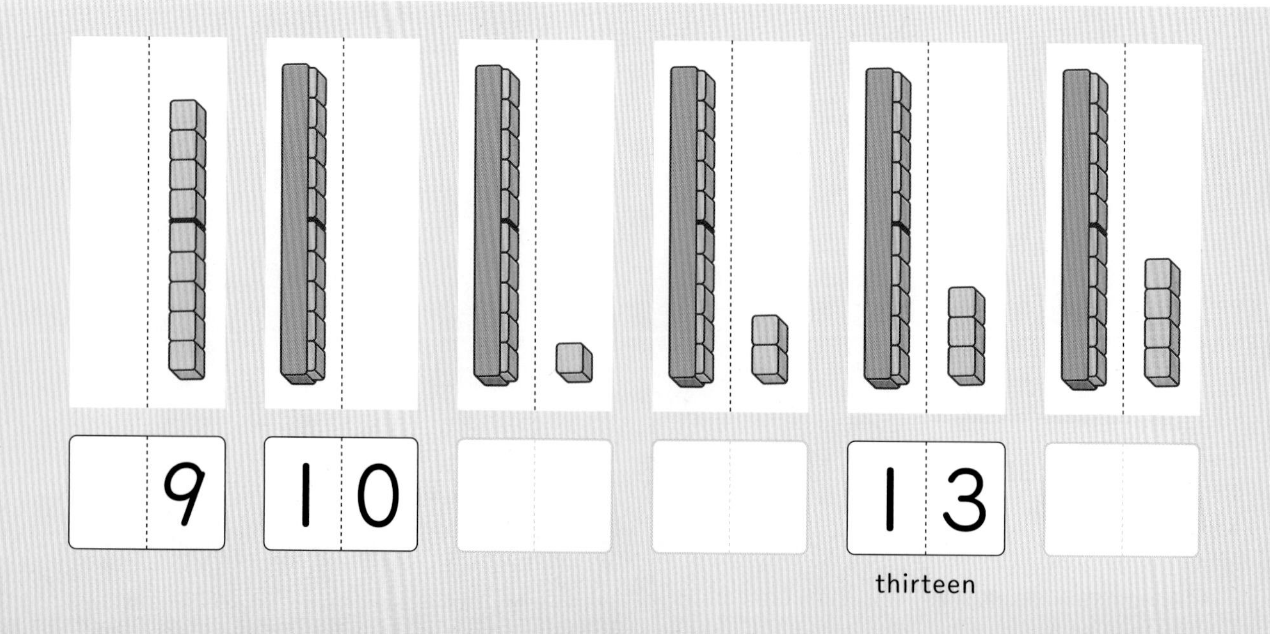

9	10			13	
				thirteen	

2 How many apples and flowers are there?

①

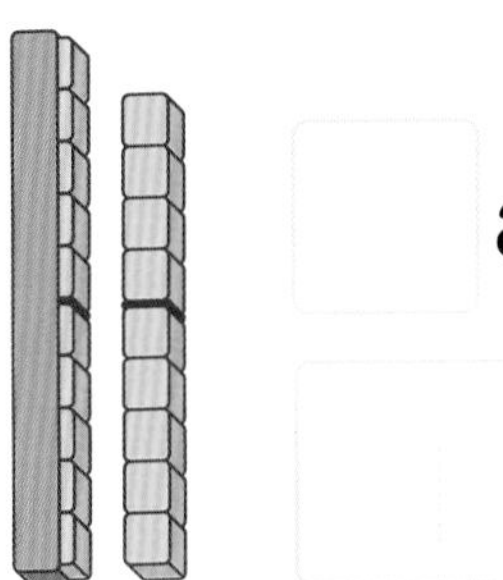

and

apples

②

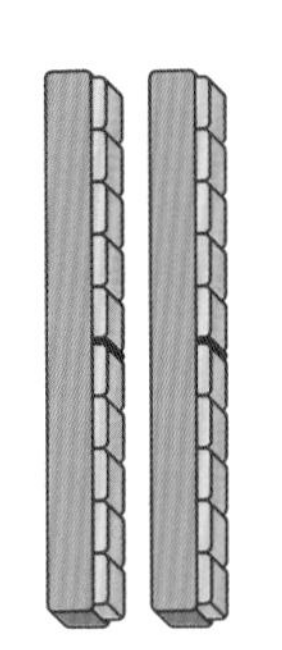

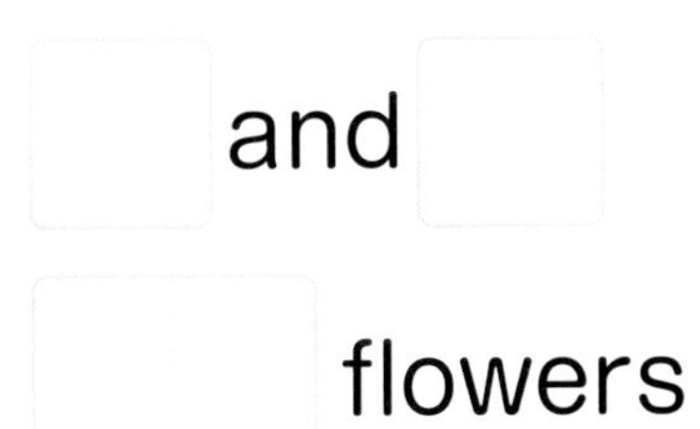

and

flowers

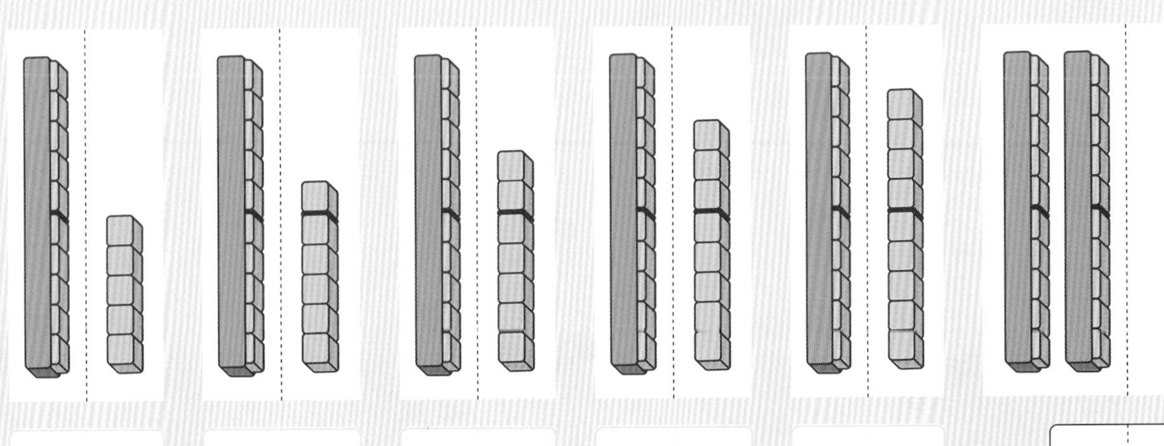

2 0

twenty

1 Let's count the things below.

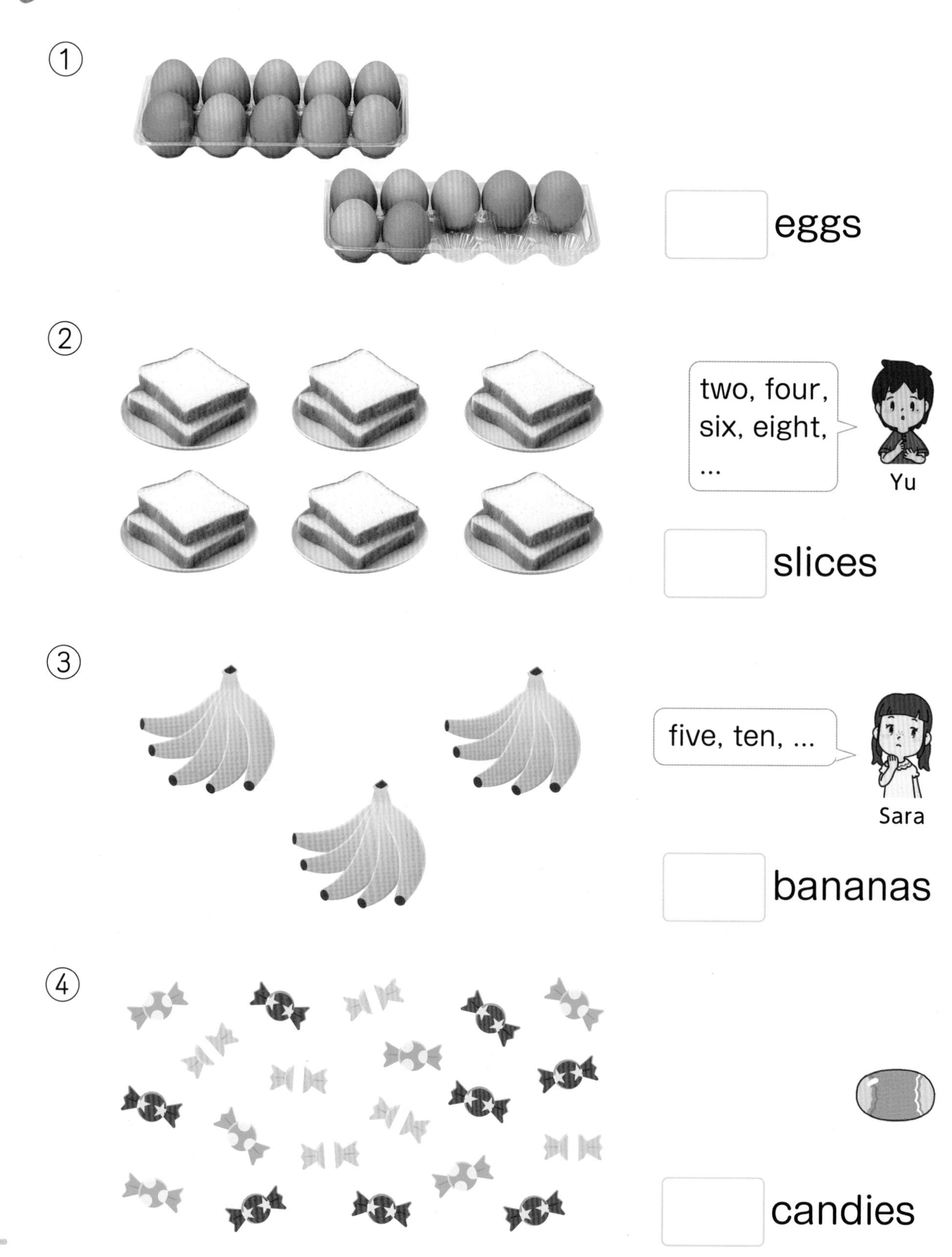

Writing numbers →

3 Let's fill in each ☐ with a number.

① 10 and 2 make ☐.

② 10 and 5 make ☐.

③ 10 and 8 make ☐.

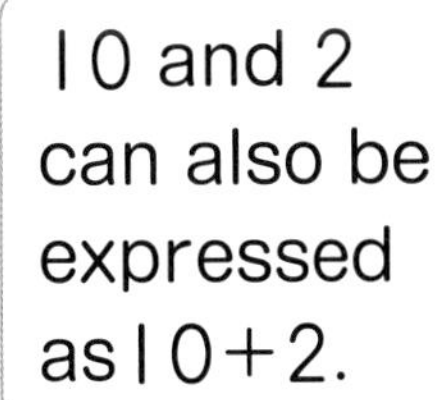

1 Let's fill in each ☐ with a number.

① 13 is ☐ and 3.

② 16 is 10 and ☐.

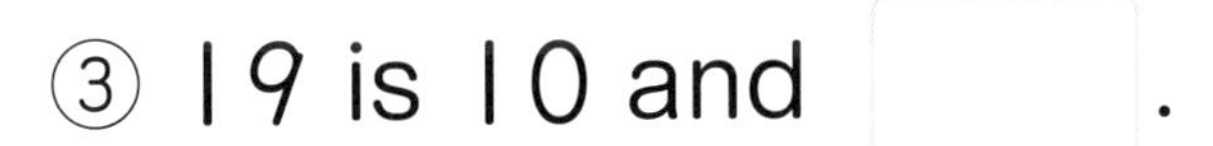

③ 19 is 10 and ☐.

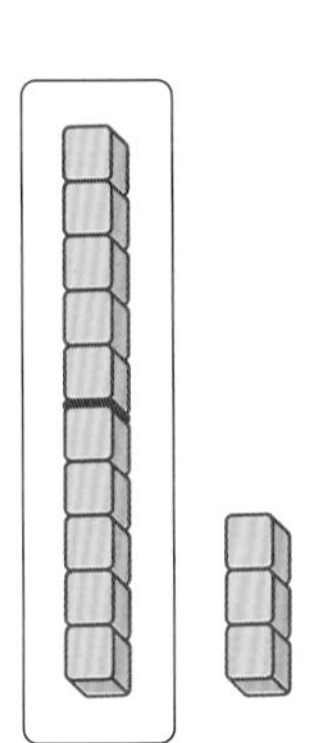

2 Let's fill in each ☐ with a number.

① 10 and 4 is ☐.

② 10 and 7 is ☐.

③ 11 is ☐ and 1.

④ 20 is 10 and ☐.

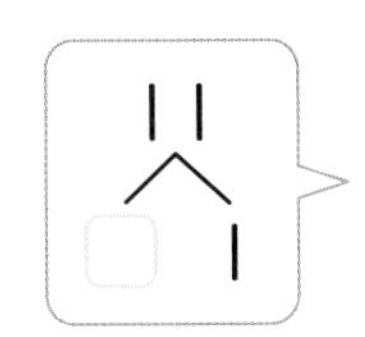

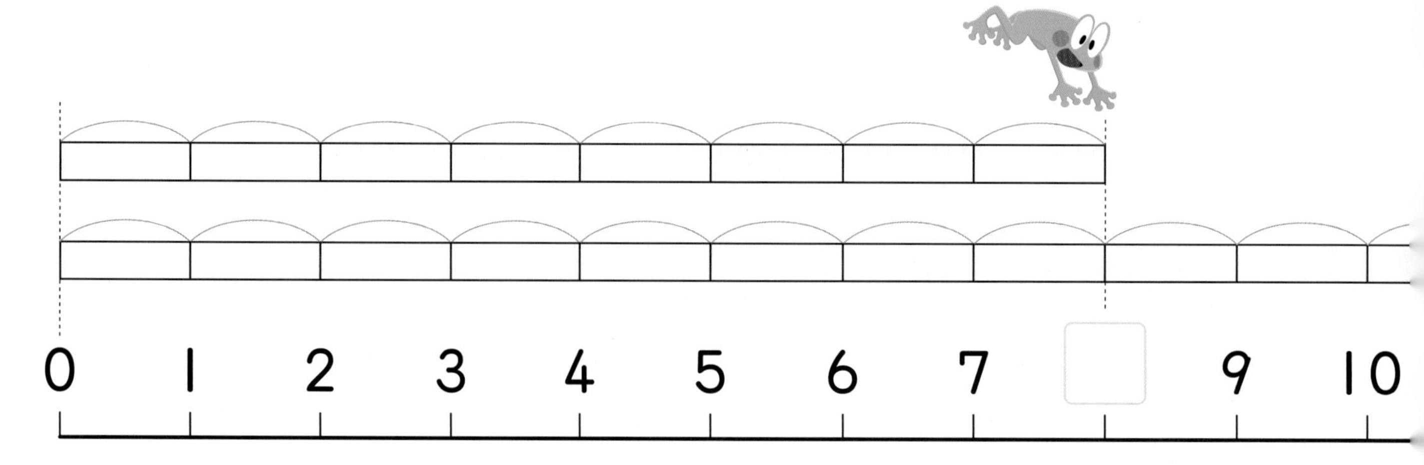

4 Let's find out the answers by using the line of numbers above.

1. Where did the  jump to?

2. Where did the  jump to?

3. Tell your friends what you found out about the line of numbers above?

As you move towards the right on the line of numbers, the number becomes...

1 Which number is larger?

① 9 or 11　　② 15 or 13

③ 14 or 17　　④ 20 or 18

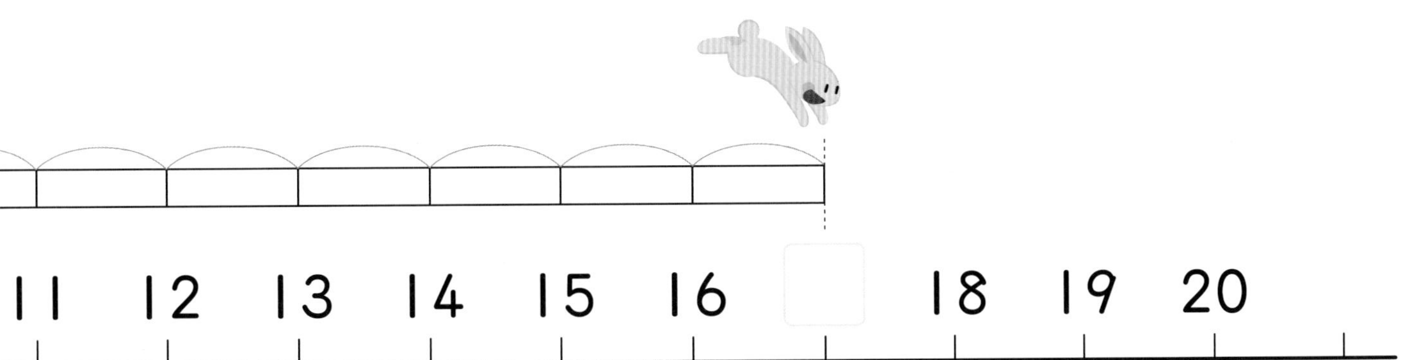

2 Let's fill in each ☐ with a number.

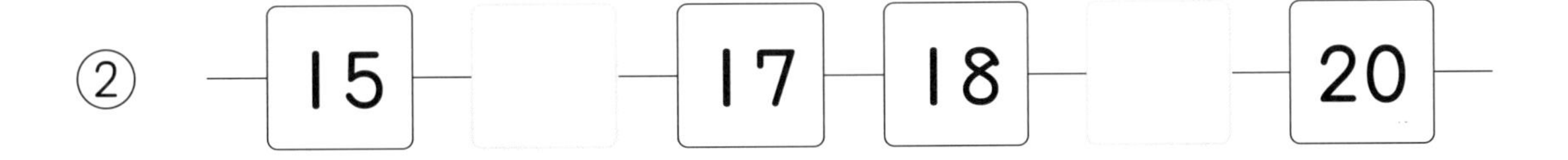

③ ☐ 19 18 17 ☐ 15

3 Let's fill in each ☐ with a number.

① 3 larger than 12 is ☐.

② 4 smaller than 18 is ☐.

? There are numbers larger than 20, right?

1

13 is 10 and 3.

Let's fill in ☐ with a number.

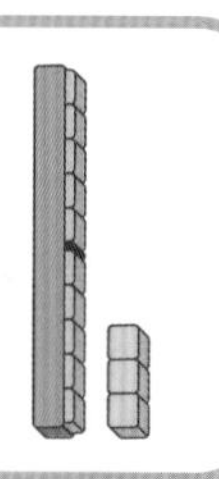

❶ The number when you add 3 to 10.

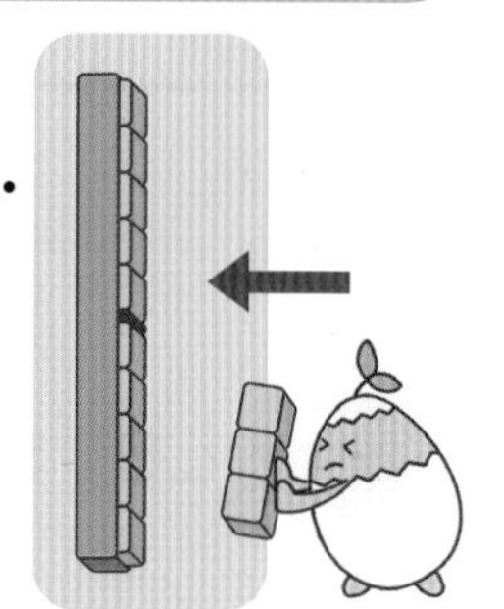

Math Sentence: $10+3=$ ☐

❷ The number when you subtract 3 from 13.

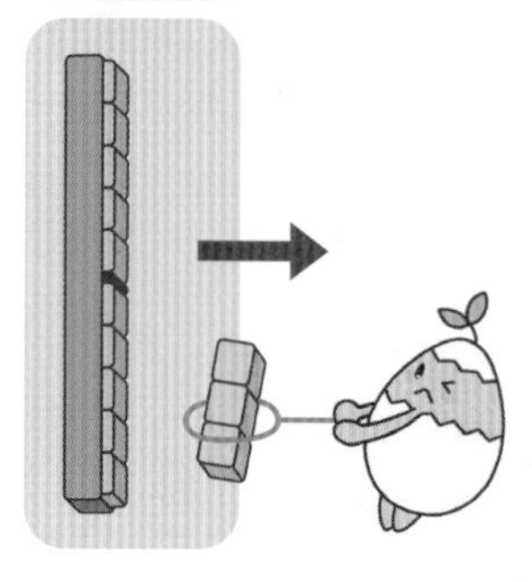

Math Sentence: $13-3=$ ☐

1 Let's operate the following.

① The number when you add 2 to 10 is ☐ .

Math Sentence: $10+2=$ ☐

② The number when you subtract 5 from 15 is ☐ .

Math Sentence: $15-5=$ ☐

③ $10+4$　④ $10+7$　⑤ $10+9$

⑥ $12-2$　⑦ $16-6$　⑧ $19-9$

2 Let's think how to find out the answers in the following problems.

① There are 12 candies. When you get 3 more candies, how many do you have altogether?

Math Sentence: ☐ + ☐ = ☐

Answer: ☐ candies

10 and 2 make 12. We add 3 to it so...

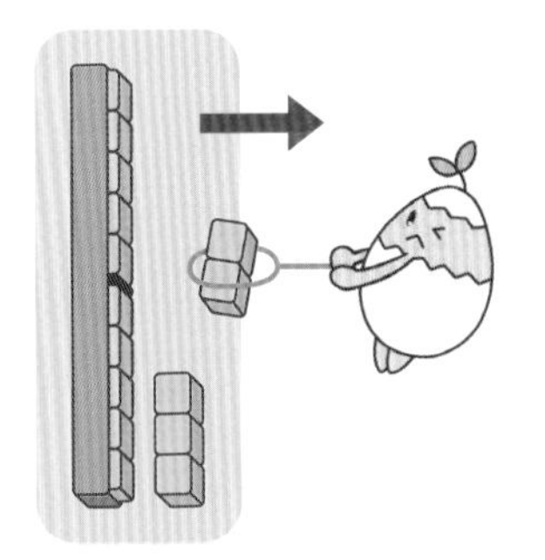

② There are 15 tomatoes. When you eat 2 tomatoes, how many are left?

Math Sentence: ☐ − ☐ = ☐

Answer: ☐ tomatoes

1 Let's operate the following.

① $11 + 4$ ② $16 + 3$ ③ $12 + 6$

④ $14 - 1$ ⑤ $17 - 2$ ⑥ $18 - 5$

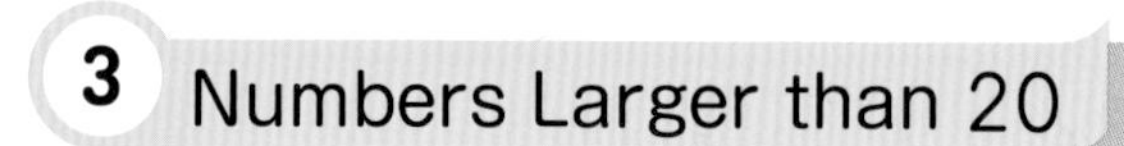

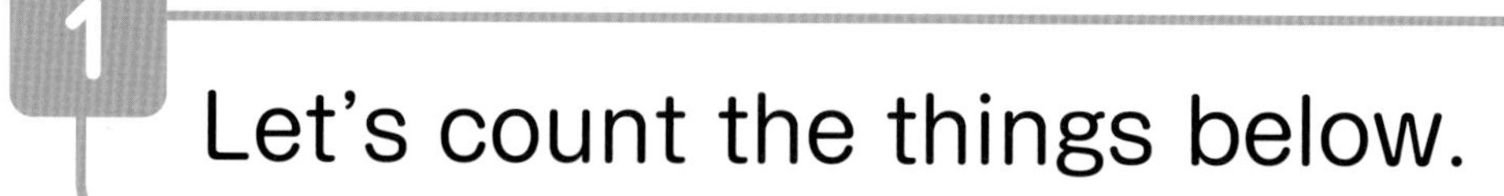

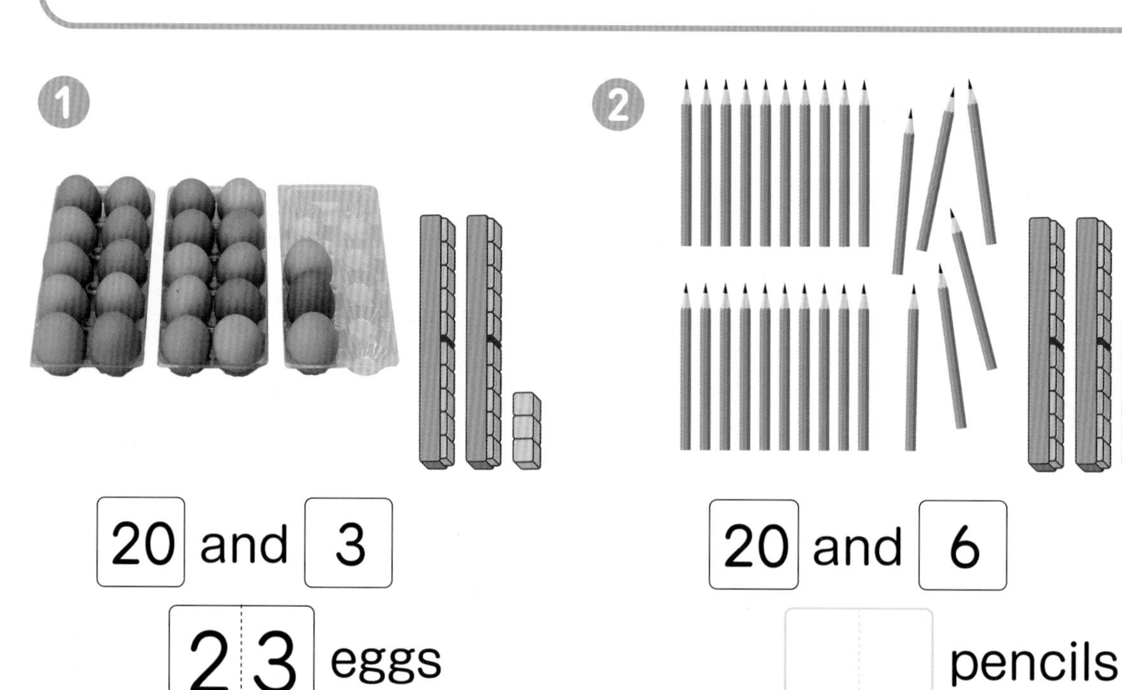

① 20 and 3

23 eggs

twenty-three

② 20 and 6

☐☐ pencils

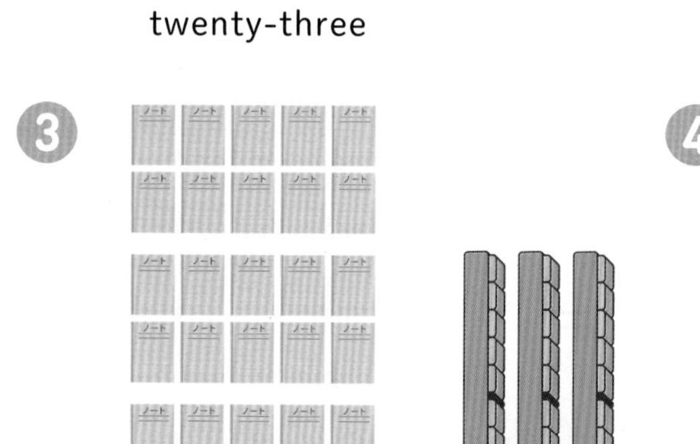

③ 3 sets of 10

☐☐ books

thirty

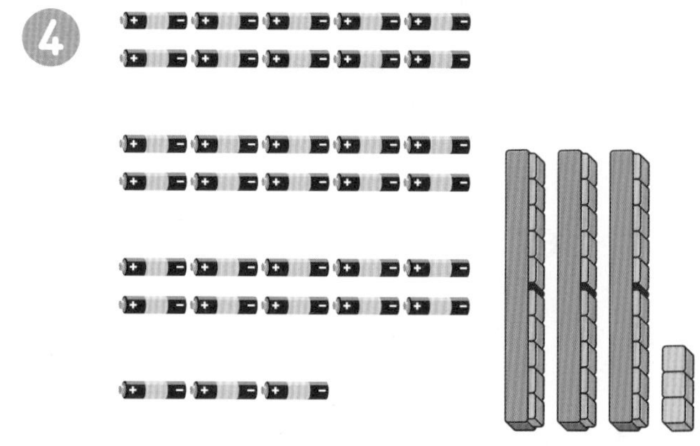

④ 30 and 3

☐☐ batteries

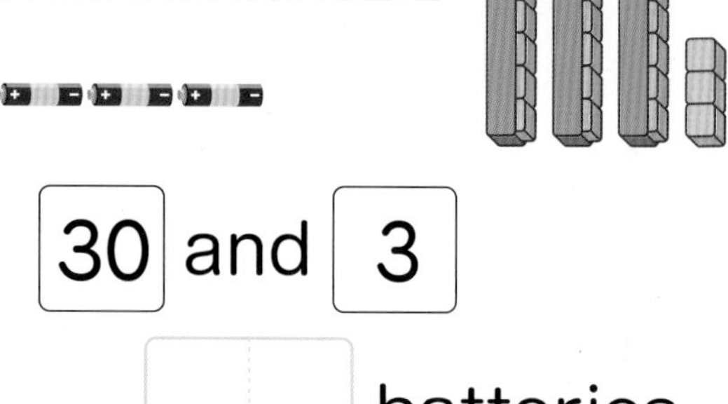

1 Let's count the number of blocks.

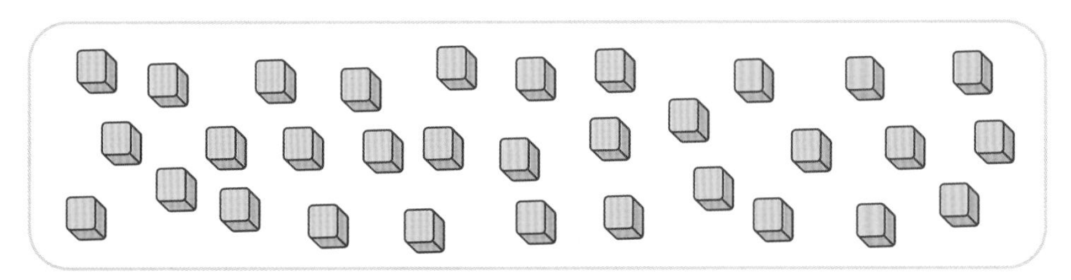

I can make sets of 10 blocks...

☐ blocks

3 sets of 10 blocks and 2 single blocks

thirty two

thirty-two

Way to see and think

It is easier to count by making sets of 10.

2 Let's read the calendar.

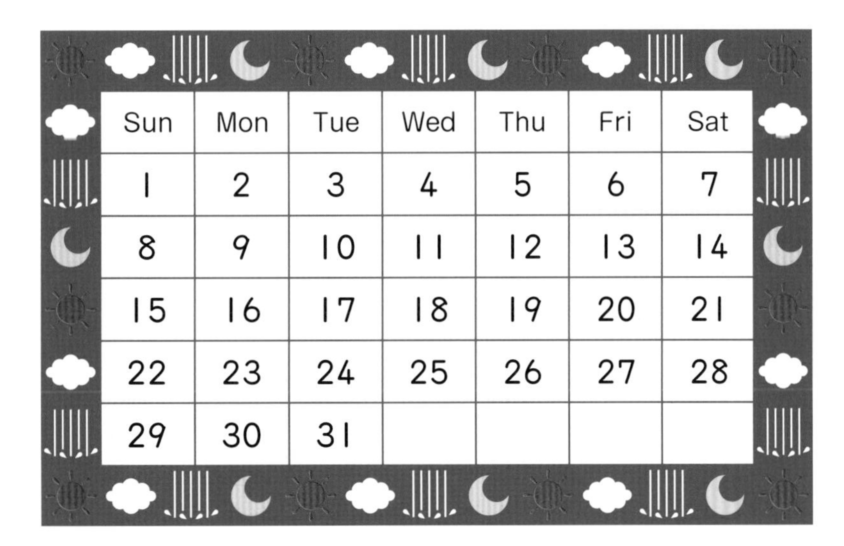

Sun	Mon	Tue	Wed	Thu	Fri	Sat
1	2	3	4	5	6	7
8	9	10	11	12	13	14
15	16	17	18	19	20	21
22	23	24	25	26	27	28
29	30	31				

? Are there larger numbers?

Supplementary Problems → p.94

8 Time (1)

1 Let's look at the pictures below and tell a story. ▷

It's our daily life.

Telling Time

The short hand is 8 and the long hand is 12, so it's 8 o'clock.

The short hand is between 8 and 9, and the long hand is 6, so it's half past 8.

Using clocks →

2 Let's show the time by moving the hands of the clock.

① 11 o'clock

② half past 3

▶1 Which clock shows 5 thirty?

① 

② 

③ 

Supplementary Problems → p.96

More Math!

[Supplementary Problems]

[Let's deepen.]

Numbers up to 10

→ pp.6 ~ 23

1 Let's write the number of dots.

① ② ③ ④

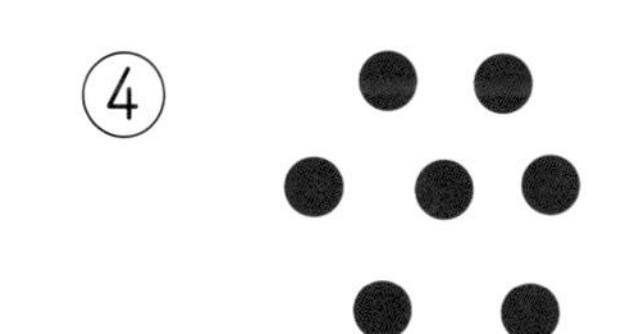

2 Which is more?

① ②

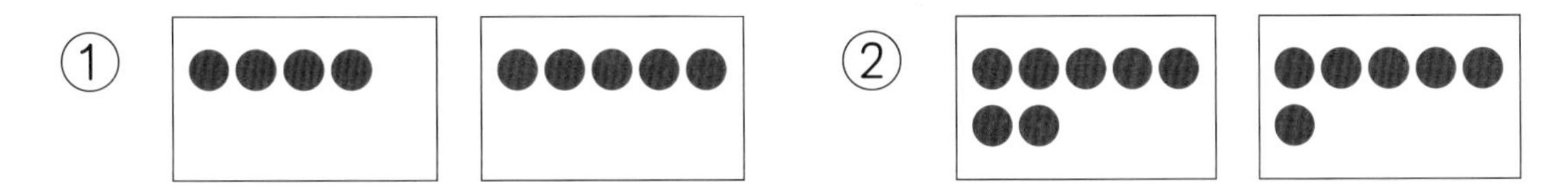

3 Which number is larger? Circle the larger number.

① ② ③ ④

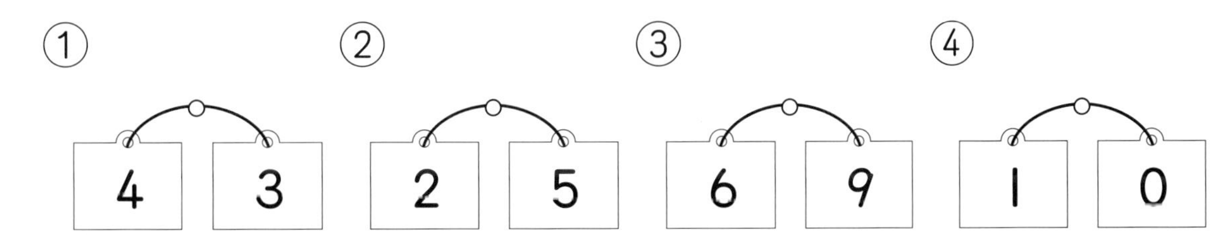

4 Let's fill in each ☐ with a number.

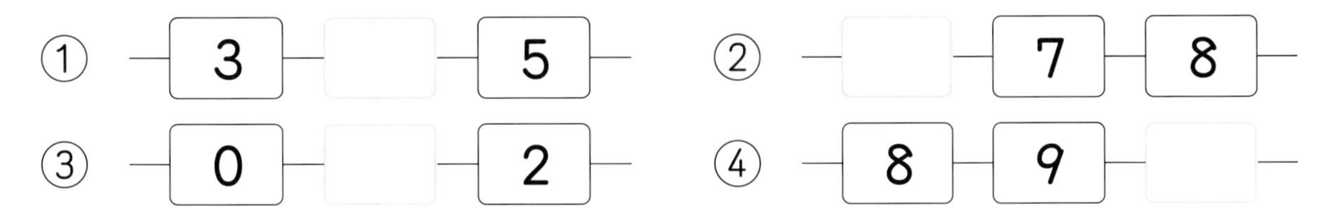

2 Decomposing and Composing Numbers

→ pp.24 ～ 31

1 Let's fill in each ☐ with a number.

① 5 is 2 and ☐.

② 5 is 4 and ☐.

③ 6 is 2 and ☐.

④ 6 is 3 and ☐.

⑤ 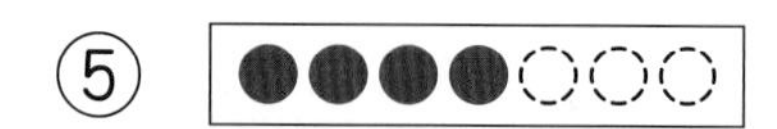 7 is 4 and ☐.

⑥ 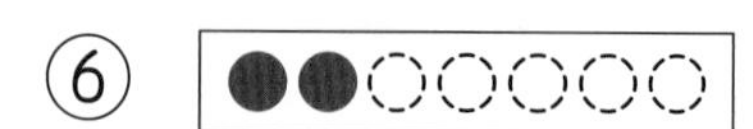 7 is 2 and ☐.

⑦  7 is 6 and ☐.

2 8 and 9 each are decomposed into 2 numbers.
Let's write the number in the ☐.

① 8 → 3 and ☐

② 8 → ☐ and 2

③ 8 → 1 and ☐

④ 8 → ☐ and 4

⑤ 9 → 7 and ☐

⑥ 9 → ☐ and 6

⑦ 9 → ☐ and 5

⑧ 9 → 8 and ☐

3 Let's connect the two numbers which make 10.

3　6　8　5　1

5　2　7　9　4

3 Ordinal Numbers

→ pp.32 ~ 35

1 Let's color it.

① The first 3 circles from the left.

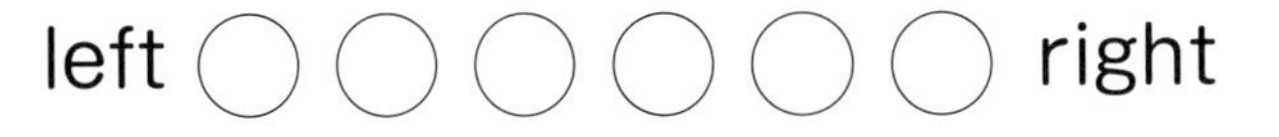

② The 3rd circle from the left.

left ○○○○○○ right

③ The first 2 circles from the right.

left ○○○○○○ right

④ The 2nd circle from the right.

left ○○○○○○ right

2 Children are lined up.

① Who is the 4th child from the front?

② What is the position of John from the front?

③ Who is the 2nd child from the back?

④ What is the position of Mei from the back?

⑤ Who is the 2nd child from the front? What is his/her position from the back?

4 How Many: Altogether and Increase

→ pp.36 ～ 53

1 There are 2 red flowers and 3 white flowers. How many flowers are there altogether?

Math Sentence: ☐　　Answer: ☐ flowers

2 There are 6 cars at the parking lot. When 3 more cars come into the parking lot, how many cars are there altogether?

Math Sentence: ☐　　Answer: ☐ cars

3 Let's do addition.

① $3+2$	② $1+1$	③ $2+1$
④ $1+4$	⑤ $1+3$	⑥ $2+5$
⑦ $7+1$	⑧ $3+6$	⑨ $5+3$
⑩ $1+8$	⑪ $4+4$	⑫ $2+7$
⑬ $3+5$	⑭ $2+8$	⑮ $1+9$
⑯ $3+7$	⑰ $5+5$	⑱ $4+6$
⑲ $7+3$	⑳ $8+0$	㉑ $3+0$
㉒ $0+2$	㉓ $0+7$	㉔ $0+0$

5 How Many: Left and Difference

→ pp.54 ~ 69

1 There are 7 goldfish. You take away 2 goldfish. How many goldfish are left?

Math Sentence: ☐ Answer: ☐ goldfish

2 There are 6 oranges and 4 apples. How many is the difference?

Math Sentence: ☐ Answer: ☐ pieces

3 There are 5 boys and 9 girls. Which is more and by how many?

Math Sentence: ☐

Answer: There are ☐ more ☐.

4 Let's do subtraction.

① $4-3$	② $9-3$	③ $8-5$
④ $6-1$	⑤ $7-3$	⑥ $3-2$
⑦ $6-5$	⑧ $8-6$	⑨ $6-3$
⑩ $10-5$	⑪ $10-3$	⑫ $10-10$
⑬ $1-1$	⑭ $4-0$	⑮ $2-0$

7

Numbers Larger than 10

→ pp.74 ～ 85

1 Let's count the number of blocks.

① ② ③

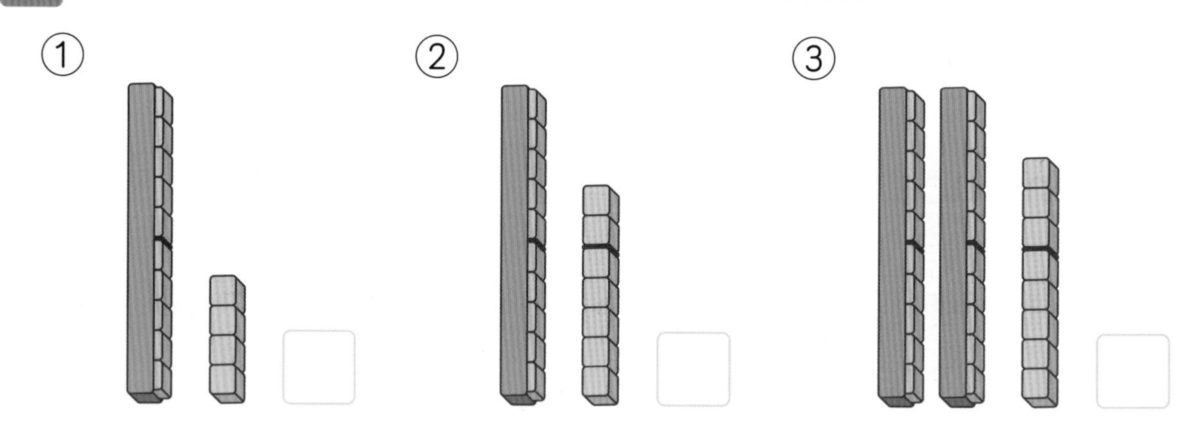

2 Let's fill in each ☐ with a number.

① 10 and 3 make ☐.

② 10 and 6 make ☐.

③ 10 and ☐ make 11.

④ 10 and ☐ make 20.

⑤ 15 is 10 and ☐.

⑥ 27 is 20 and ☐.

3 Which number is larger? Circle the larger number.

①

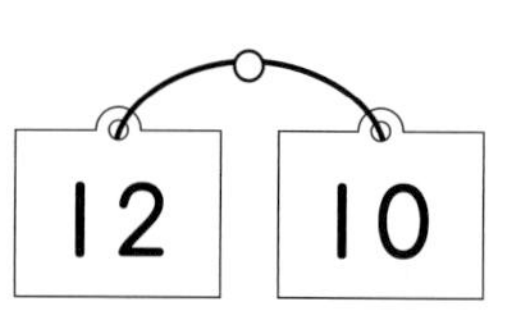

②

③

4 Let's fill in each ☐ with a number.

5 Let's find out the answers.

① $10+5$	② $10+8$	③ $14-4$
④ $18-8$	⑤ $12+5$	⑥ $11+6$
⑦ $14+4$	⑧ $17+2$	⑨ $13-2$
⑩ $15-4$	⑪ $17-3$	⑫ $19-5$

6 There are 11 sheets of origami paper. 8 more sheets are given. How many sheets are there altogether?

Math Sentence: ☐ Answer: ☐ sheets

7 There were 16 candies. You gave your friend 5 candies. How many candies are left?

Math Sentence: ☐ Answer: ☐ candies

8 Time (1)

→ pp.86 ～ 87

1 What time is it?

①

②

③

④

2 Draw the long hand on the clock and show the time.

① 5 o'clock

② 11 o'clock

③ two-thirty

④ nine-thirty

Animal Quiz

There are some animals lining up.
Which animal does Yu like?

① It is a very cute animal.
② It is not the 4th animal from the right.
③ It is one of the three animals from the left.
④ It is not next to the mouse.

Rock, Paper, Scissors Game

Let's play rock, paper, scissors game.

If you win with paper, 2 steps forward!

If you win with rock, 4 steps forward!

If you win with scissors, 6 steps forward!

Akari

I won twice, and moved 6 steps forward. Which one did I win with?

Haruto

I want to move 10 steps forward. With which one and by how many should I win?

If I win with scissors, I can move 6 steps forward. How many more steps do I need to move 10 steps forward?

6. How Many in Each Group: Activity Cards

→ To be used in Page 72.
Please cut these out for use.

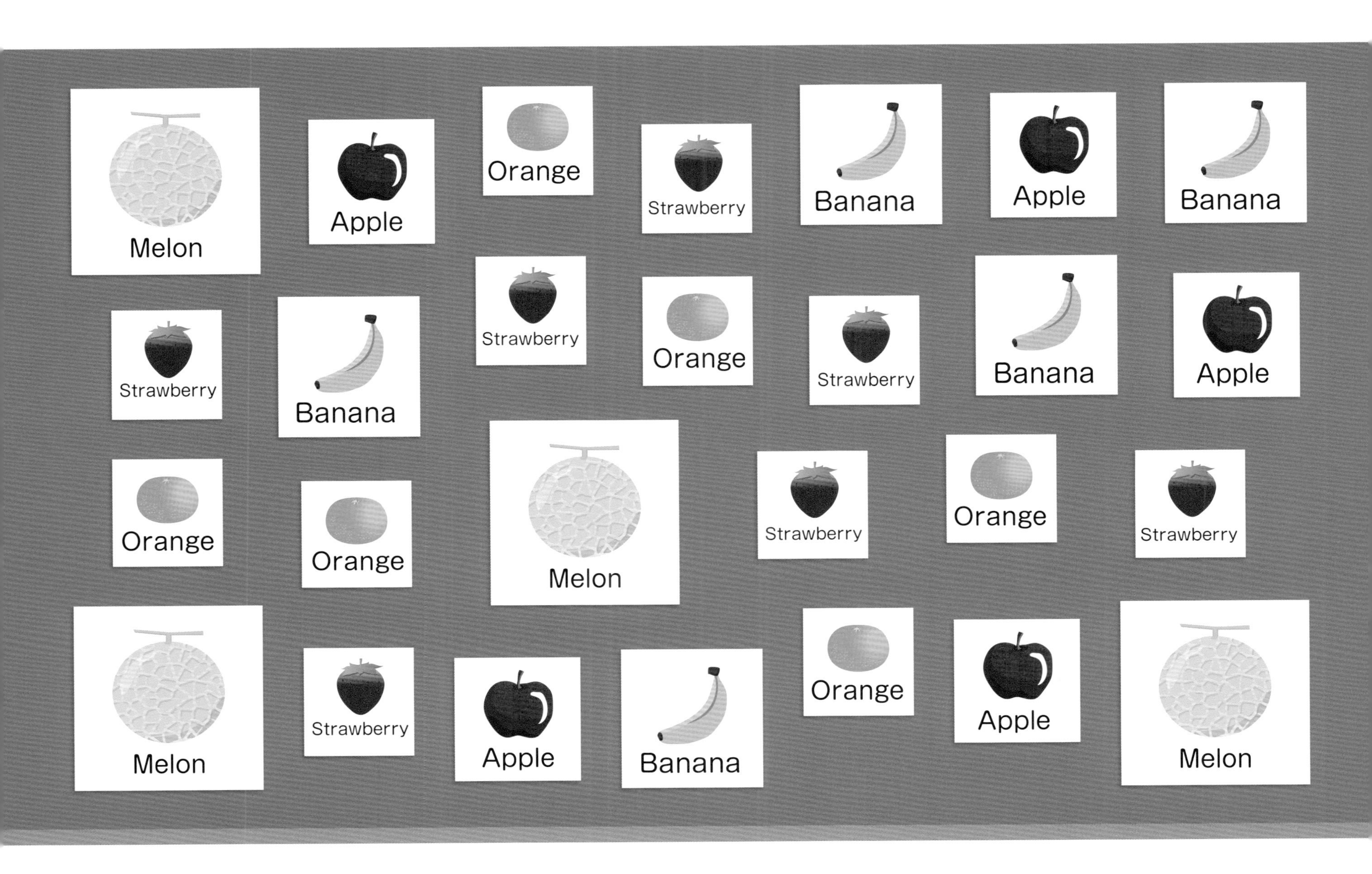

Memo

Memo

Memo

Editors of Original Japanese Edition

[Head of Editors]

Shin Hitotsumatsu (Kyoto University), Yoshio Okada (Hiroshima University)

[Supervising Editors]

Toshiyuki Akai (Hyogo University), Toshikazu Ikeda (Yokohama National University), Shunji Kurosawa (The former Rikkyo University), Hiroshi Tanaka (The former Tsukuba Univ. Elementary School), Kosho Masaki (The former Kokugakuin Tochigi Junior College), Yasushi Yanase (Tamagawa University)

[Editors]

Shoji Aoyama (Tsukuba Univ. Elementary School), Kengo Ishihama (Showa Gakuin Elementary School), Hiroshi Imazaki (Hiroshima Bunkyo University), Atsumi Ueda (Hiroshima University), Tetsuro Uemura (Kagoshima University), Yoshihiro Echigo (Tokyo Gakugei Univ. Setagaya Elementary School), Hisao Oikawa (Yamato University), Hironori Osawa (Yamagata University Graduate School), Tomoyoshi Owada (Shizuoka University), Nobuhiro Ozaki (Seikei Elementary School), Masahiko Ozaki (Kansai Univ. Elementary School), Kentaro Ono (Musashino University), Hiroshi Kazama (Fukui University), Michihiro Kawasaki (Oita University), Miho Kawasaki (Shizuoka University), Yoshiko Kambe (Tokai University), Yukio Kinoshita (Kwansei Gakuin Elementary School), Tomoko Kimura (Tamon Elementary School, Setagaya City), Satoshi Kusaka (Naruto University of Education), Kensuke Kubota (Naruo Higashi Elementary School, Nishinomiya City), Itsushi Kuramitsu (The former University of the Ryukyus), Maiko Kochi (Kounan Elementary School, Toshima City), Chihiro Kozuki (Daiyon. Elementary School, Hino City), Goto Manabu (Hakuoh University), Michihiro Goto (Tokyo Gakugei Univ. Oizumi Elementary School), Hidenori Kobayashi (Hiroshima Univ. Shinonome Elementary School), Akira Saito (Shibata Gakuen University Graduate School), Masahiko Sakamoto (The former Tokoha University Graduate School), Junichi Sato (Kunitachigakuen Elementary School), Hisatsugu Shimizu (Keio Yochisha Elementary School), Ryo Shoda (Seikei University), Masaaki Sugihara (University of the Sacred Heart, Tokyo), Jun Suzuki (Gakushuin Primary School), Shigeki Takazawa (Shiga University), Chitoshi Takeo (Nanzan Primary School), Hidemi Tanaka (Tsukuba Univ. Elementary School), Toshiyuki Nakata (Tsukuba Univ. Elementary School), Hirokazu Nagashima (Daisan. Elementary School, Kokubunji City), Minako Nagata (Futaba Primary School), Kiyoto Nagama (Hiyagon Elementary School, Okinawa City), Satoshi Natsusaka (Tsukuba Univ. Elementary School), Izumi Nishitani (Gunma University), Kazuhiko Nunokawa (Joetsu University of Education), Shunichi Nomura (Waseda University), Mantaro Higuchi (Kori Nevers Gakuin Elementary School), Satoshi Hirakawa (Showa Gakuin Elementary School), Kenta Maeda (Keio Yokohama Elementary School), Hiroyuki Masukawa (University of the Sacred Heart, Tokyo), Shoichiro Machida (Saitama University), Keiko Matsui (Hasuike Elementary School, Harima Town), Katsunori Matsuoka (Naragakuen University), Yasunari Matsuoka (Matsushima Elementary School, Naha City), Yoichi Matsuzawa (Joetsu University of Education), Satoshi Matsumura (Fuji Women's University), Takatoshi Matsumura (Tokoha University), Kentaro Maruyama (Yokohama National Univ. Kamakura Elementary School), Kazuhiko Miyagi (Homei Elementary School Affiliated with J.W.U), Aki Murata (University of California, Berkeley), Takafumi Morimoto (Tsukuba Univ. Elementary School), Yoshihiko Moriya (The former Kunitachigakuen Elementary School), Junichi Yamamoto (Oi Elementary School, Oi Town), Yoshikazu Yamamoto (Showa Gakuin Elementary School), Shinya Wada (Kagoshima University), Keiko Watanabe (Shiga University)

[Reviser]

Katsuto Enomoto (The former Harara Elementary School, Kagoshima City)

[Reviser of Special Needs Education and Universal Design]

Yoshihiro Tanaka (Teikyo Heisei University)

[Cover]

Photo : Tatsuya Tanaka (MINIATURE LIFE)

Design : Ai Aso (ADDIX)

[Text]

Design : Ai Aso, Takayuki Ikebe, Hanako Morisako, Miho Kikuma, Sawako Takahashi, Katsuya Imamura, Satoko Okutsu, Kazumi Sakaguchi, Risa Sakemoto (ADDIX), Ayaka Ikebe

[Illustrations]

Lico, Gami, Shiho Sakai, Tomoko Shintani, Akira Sugiura, Nanami Sumimoto, Ryoko Totsuka, Kinue Naganawa, Yoshie Hayashi, B-rise

[Photo・Observation data]

Mihara Kindergarten, Showa Gakuin Elementary School, ABstudio, Ken Ogawa, DEPO LABO, Ritmo